SPATIAL DIFFUSION

RICHARD MORRILL
Department of Geography
University of Washington

GARY L. GAILE
Department of Geography
University of Colorado

GRANT IAN THRALL
Department of Geography
University of Florida

SAGE PUBLICATIONS
The Publishers of Professional Social Science
Newbury Park Beverly Hills London New Delhi

This book is dedicated to
Torsten Hägerstrand

Copyright © 1988 by Sage Publication, Inc.

All rights reserved. No part of this book may be reproduced or utilized in any form or by any means, electronic or mechanical, including photocopying, recording, or by any information storage and retrieval system, without permission in writing from the publisher.

For information address:

SAGE Publications, Inc.
2111 West Hillcrest Drive
Newbury Park, California 91320

SAGE Publications Inc.
275 South Beverly Drive
Beverly Hills
California 90212

SAGE Publications Ltd.
28 Banner Street
London EC1Y 8QE
England

SAGE PUBLICATIONS India Pvt. Ltd.
M-32 Market
Greater Kailash I
New Delhi 110 048 India

Printed in the United States of America

International Standard Book Number 0-8039-2852-1
0-8039-2684-7

Library of Congress Catalog Card No. L.C. 87-062683

SECOND PRINTING, 1988

CONTENTS

Series Editor's Introduction 5

1. **Introduction: Diffusion as a Space-Time Process** 7
   Importance of Space-Time Diffusion 7
   The Nature and Manner of Spatial Diffusion 8
   Diffusion: A Descriptive Typology 10
   Illustrations of Diffusion Phenomena 14

2. **Nongeographic Diffusion Literature** 16
   Physical Sciences and Mathematics 16
   Biological Sciences 17
   Sociology 18
   Economics 19
   History 19
   Anthropology 19
   Conclusion 20

3. **Development of the Basic Concept of Diffusion as a Spatial Process** 21
   Introduction to the Hägerstrand Model 23
   Theory and Models 25
   The Empirical Studies 29
   The Monte Carlo Model 30
   Testing the Model 32
   Conclusion 34

4. **Spatial Diffusion Processes: Post-Hägerstrand Developments** 34
   Elaboration of the Hägerstrand Model 35
   Resistance 36
   Barriers 37
   Barriers in Post-Hägerstrand Wave Theories 40
   Variable Mean Information Fields 41
   Measures of Some Characteristics
       of Spatial Diffusion 42
   Stochastic and Deterministic Conceptions
       of Diffusion 45
   Hierarchical Diffusion 47
   Spatial Diffusion and Spatial Interaction 49

The Role of the Propagator:
      The Market and Infrastructure Perspective   52
   Diffusion and Development:
      Diffusion as a Planning Tool   55
   Conclusion   56

5. **Mathematical Approaches to Diffusion Modeling**   **57**
   Introduction   57
   Stochastic and Deterministic Variants
      of the Hägerstrand Model   57
   General Interaction Diffusion Models
      and Critical Measures   62
   Epidemiology Models   63
   General Space-Time Models   64
   Spatial-Temporal Models   65
   Measurement and Evaluation
      of Expected Diffusion Behavior   72
   Conclusion   73

6. **Present Status, Needs, and Developments**   **73**
   Needs and Future Trends   78

References   **80**

About the Authors   **85**

SERIES EDITOR'S INTRODUCTION

This volume is about how we come to have the culture and ideas we have. Most social and economic change is a direct consequence of the diffusion of some idea or phenomenon. Ideas become diffused through society in a regular manner, and because of this regularity their diffusion can often be analyzed and even predicted. The same analytical framework that can be applied to describe and predict the spread of some cultural or human phenomenon, such as political turmoil, can also be applied to an analysis of the spread of disease.

The research on diffusion begins with an analysis of the origin of the phenomenon. What are the characteristics of the *person* that comes up with the new ideas. What are characteristics of the *place* from which the phenomenon eminates. And then what typifies the successive persons and places that adopt the phenomenon. Case studies of such people and places make for fascinating reading, but it is the general theory that binds together the circumstances that surround the innovative people and places of origin. The general theories of diffusion are presented in this volume, along with supporting examples.

The innovative person cannot be understood in isolation from the rest of society, for it is that society that judges whether or not to adopt or reject the innovation. And it takes *time* for the phenomenon to be accepted by the various sectors of society.

Places cannot be viewed in isolation from one another for it is the linkage of information flows that determine when and to what extent the new phenomenon can be adopted in other places: closer and larger places often receive the stimulus for the new phenomenon before places that are small and remote. Therefore, a general theory of diffusion cannot be complete without an understanding of the role of *space* in the diffusion process.

This volume chronicles the evolution of ideas for analyzing, simulating, and forecasting the diffusion of phenomena. This chronicle is done knowing that the direction the research community would take has been toward a synthesis of the roles of time and space, how they interdependently govern the diffusion of phenomena, and how such an understanding could be used to enhance the scientific predictability of diffusion in a wide array of contexts.

Spatial diffusion is defined and described in the introductory chapter. A brief review of salient diffusion research in a variety of academic disciplines is presented in the second chapter. Both chapters include example applications of the major diffusion theories.

A review of Torsten Hagerstrand's seminal research on the geographic processes of diffusion is presented in the third chapter. An elementary step-by-step example of the Hagerstrand model is included, and then refined by adding improvements of the theoretical arguments so that a better approximation to reality can be achieved by the model. The fourth chapter details the post-Hagerstrand contributions to spatial diffusion theory.

The more recent literature on spatial diffusion has focused on issues of how to improve the precision and robustness of the increasingly mathematically complex diffusion analysis; these developments are reviewed in Chapter 5. The chapter ends with a discussion of the present research thrust to integrate temporal and spatial models of diffusion. The sixth chapter is a forum for reflection upon the overall importance of the past accomplishments and speculation on the future of diffusion research.

A great array of academic disciplines other than geography have also contributed to the research on diffusion, including epidemiology, mathematics, physics, sociology, anthropology, planning, management science and marketing, economics, and history. Because much innovation comes from borrowing, students and professionals in each of these diverse areas should find this volume on spatial diffusion of significant value.

—*Grant Ian Thrall*
Series Editor

SPATIAL DIFFUSION

RICHARD MORRILL
University of Washington
GARY L. GAILE
University of Colorado/Harvard University
GRANT IAN THRALL
University of Florida

1. INTRODUCTION: DIFFUSION AS A SPACE-TIME PROCESS

Importance of Space-Time Diffusion

How do we come by information and adopt new ideas? What is the pattern of disease transmission, and the spread of animal and plant life? These are examples of diffusion phenomena and processes. The questions can be rephrased as the general query: How do phenomena become diffused? Although there are certain aspects that make each phenomenon unique as it is diffused through space, at the same time all phenomena that become diffused share general spatial patterns and processes. This book is a review of the general theories of spatial diffusion.

Spatial diffusion is the process by which behavior or characteristics of the landscape change as a result of what happens *elsewhere earlier*. Spatial diffusion is the spread of the phenomenon, over space and timed, from limited origins.

Diffusion processes are common in nature. A dramatic example is the sudden spread of material and shock waves from a volcanic explosion such as the 1980 eruption of Mount Saint Helens in the state of Washington; more slowly, but still a diffusion process, is the subsequent reestablishment of plant and animal life there. Hence the diffusion process can occur suddenly, or even take millenia.

On a global scale, and over the millenia of human experience, spatial diffusion encompasses the spread of the settlement of humanity itself. The spatial diffusion of technological and intellectual characteristics of civilization includes the spread of food and clothing, implements and techniques of war, style of shelter, and urbanization. Conflict between people often results as these agents of change become diffused. On a regional scale, spatial diffusion includes the spread of diseases and epidemics, the territorial expansion of individual cities, and the expansion of ethnic neighborhoods. Diffusion is therefore a spatial process that can transform the human and physical landscape.

The Nature and Manner of Spatial Diffusion

The key elements of spatial diffusion are phenomenon and spread: some phenomenon is somehow brought into existence; that phenomenon moves or is spread beyond the origin to alter, even temporarily, the character of other places.

The study of spatial diffusion emphasizes the manner of spatial movement and spatial change. However, the process of the creation of the phenomenon that may become diffuse is also important, but often difficult to explain and identify.

What can be the characteristics of the phenomenon that is being diffused? The phenomenon can be material such as human settlement; it can be immaterial such as an idea or mode of behavior. The phenomenon must have a real place of origin; it can be generated through a physical process such as that of the volcano explosion example, or by human decision. The phenomenon must be transferable; if it is an idea, the phenomenon must be adoptable or adaptable by oneself or others who in turn become agents through which further spread occurs. For example, a phenomenon may be a contagious disease, and is created when an individual becomes infected. The disease can be spread or diffused if the individual contacts others, and the nature of the spread reflects the geographic and temporal pattern of the individual's contacts.

Diffusion does take place; but, by what mechanism does it occur? Besides the propagation of the phenomenon at the origin, there must be agents that have the means to effect transfer of the phenomenon, and places (or beings at places) that can receive the phenomenon. Agents can be inanimate such as wind and water; agents can be animate such as people who transfer phenomena in person or by indirect communication. Diffusion can be modified by costs and quality of the paths over which the phenomenon moves, by the attractiveness of the phenomenon, and by the ability of the surrounding territory or its people to absorb the phenomenon. For instance, a disease cannot spread if the remainder of the population is immune to the disease; the likelihood that a new trait will be adopted is lessened if the adopters do not view the trait as useful or profitable.

For a phenomenon to be considered as being an example of a diffusion process, it must be shown to have originated at a certain place and at a particular point in time. For example, plot over time the cumulative proportion of population in a region that has purchased a microcomputer: if $x_1\%$ of the population make the purchase in time period 1, and $x_2\%$ of the population make the purchase in time period 2, then the cumulative percentage of population that have bought the good at the end of time period 2 is $x_1\% + x_2\%$. The cumulative percentage at the end of the nth time period is $x_1\% + x_2\% + \ldots x_n\%$. Most often, the result of such experiments reveals an S-shaped pattern (Figure 1.1). Even if the rate of making contacts among individuals remains constant, the cumulative proportions will first rise exponentially, as the early buyers spread through a large "susceptible" population: susceptible persons are those that could, but have not yet, bought a microcomputer. Then, the rate must taper off since the ratio of nonowners to owners necessarily falls. The tapering off is largely attributable to a lack of opportunity to make additional contacts with persons of a declining susceptible population that have not yet purchased a computer. Also, the passage of time generally diminishes the "missionary-like zeal" of early purchasers of equipment; the appeal of the phenomenon itself may diminish because it has been around for some time.

It is also useful to place spatial diffusion within the broader context of all movements. There are two broad classes of movements or flows: spatial diffusion and spatial interaction. Spatial diffusion encompasses some of those movements that alter the landscape or change behavior at locations; this is in contrast to spatial interaction, which are those flows inherent or necessary to the everyday functioning of an existing structure. Spatial interaction includes the journey to work, school, or

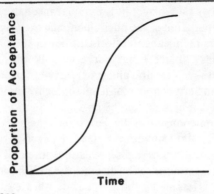

SOURCE: Morrill, 1968.

Figure 1.1 Cumulative Acceptance of an Innovation in Time (in an area)

shopping; most internal and foreign trade; vacation travel; and the material cycles of the physical world (such as the carbon cycle). However, all these forms of spatial interaction are important sources of information about traits or channels of contact with others that may influence spatial diffusion.

Diffusion: A Descriptive Typology

The nature of "spread" or movement from an "origin" implies a spatial continuity; the idea of spatial continuity is captured in the statistical concept of spatial autocorrelation: That which happens at one place is in part a function of what has already happened at nearby places. One class of diffusion processes where the spread is smooth and continuous across space is known as *contagious*, an analogy from epidemiology.

Contagious diffusion can be visualized as being like a wave spreading from a rock falling in water. The most significant characteristic of this pattern of diffusion is that a zone of the most active rate and amount of change moves outward from the origin like the crest of a wave. One of the key tests of whether diffusion occurs is whether this wave-like characteristic can be detected. The physical spread of a city into its rural hinterland is one example. There is an urban edge of the greatest intensity of new housing construction (the phenomenon); yet some new housing is added in advance of the most active edge while filling-in continues in areas largely developed.

The importance of the contagious process may become lessened as populations become concentrated in a few metropolitan areas and as greater numbers of people get information from national and even international media. In contemporary society, the process of diffusion by contagion for some phenomena has become less important. Nonetheless, space still remains as a barrier to the flow of many phenomena, thereby reserving a "global village"-like world for the future. In the future, diffusion by contagion may become more important with human settlement in space and on other planets.

It is rare for human phenomena to follow a perfect pattern of the contagious type because this is not the way that people interact. The book by King (1984) in this series outlines the theory of central places: There are relatively few large places and increasingly more small places on the landscape; small places are closer together while the largest places generally have maximum spatial dispersion. The book by Haynes and Fotheringham (1984) also in this series outlines the theory of gravity and spatial interaction models: Interaction is likely to be greater between places of larger populations and interaction is likely to be less the more remote the place is. Thus many phenomena become diffused in a *hierarchical* manner—a phenomenon may be observed first in the largest city, then jump across the landscape to the next largest, and so on. For instance, a trait may become popular in New York City, may next appear in Los Angeles, then Chicago, then Dallas, and only much later in Buffalo, even though Buffalo is closer to New York City than is Dallas. Figure 1.2b depicts the hierarchical pattern of diffusion in contrast to an ideal contagious diffusion pattern depicted in Figure 1.2a.

Distinguishing between whether the spatial pattern is hierarchical or contagious can assist in revealing whether the phenomena being diffused result from physical or human processes. Phenomena being diffused with the aid of people are more likely to follow a hierarchical pattern; it is unlikely for physical phenomena unassisted by people to follow a hierarchical pattern.

If the process of diffusion results in more agents that possess the phenomenon, such as more people with the trait, then we refer to the process as *expansion*-type diffusion. In contrast, if the number of agents with the diffusion characteristic does not increase, then we refer to the process as *relocation*-type diffusion; the agents may have merely changed spatial location, or as the trait is passed on to additional agents, it is lost in the origial agent. Figure 1.2c illustrates relocation-type diffusion. Of course, many real world diffusions display characteristics of more than one type of diffusion.

Different spatial patterns may result from a diffusion that is actively promoted, in contrast to one that is passively accepted. Diffusion that is passive depends mainly on the potential adopter, while diffusion that is

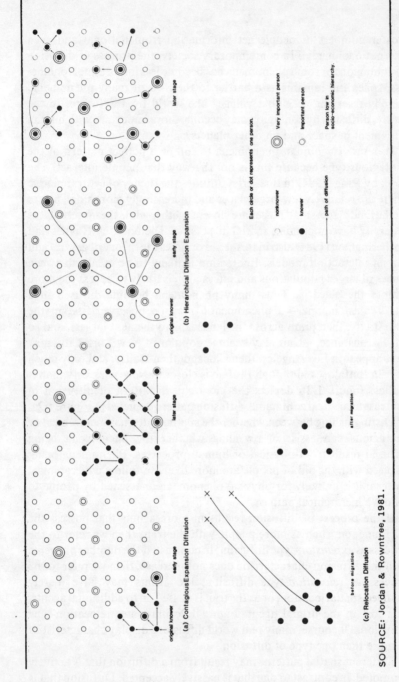

SOURCE: Jordan & Rowntree, 1981.

Figure 1.2 Idealization of Spatial Diffusion Categories

promoted can be directed by an entrepreneur or government. For example, adoption of a new product may rely on the existing structure of retail outlets (passive) or may be actively promoted by establishment of a new set of exclusive outlets. In either case, contagious or hierarchical patterns may occur; generally, a smoother pattern will appear on the landscape if the diffusion is promoted.

Two general approaches have been used to model the process of diffusion: stochastic and deterministic. A stochastic model is one in which the elements include probability. Deterministic models, in contrast, do not allow for chance. The stochastic stream of the literature suggests that an observed spatial pattern of diffusion phenomena may be the result of forces that have a random component. For example, the order in which a rumor is diffused can be considered to be a stochastic process; a perfect contagious process is unlikely since much depends on whom the carrier of the rumor chances upon first and if "susceptibles" are at home. Also, the pattern that may be interpreted by a determinist as indicative of a contagious process may also be viewed by a stochastic modeler as the consequence of random processes: one outcome of many possible alternatives.

The introduction of the random or stochastic element into the diffusion process is both one of philosophical orientation and an issue of operational modeling. On the one hand, when the diffusion process that is being modeled is at the large or macroscale, the process is often described in terms that are deterministic; all significant considerations are included in the model and no element of chance is considered. On the other hand, when the diffusion process that is being modeled is at the small or microscale, the process is often described in stochastic terms; for illustration, one has a certain probability of encountering the agent possessing the trait to be diffused. It may be logically possible to identify all the events in minutia that were responsible for the encounter and thereby construct a deterministic model, but in practice it would be unlikely that the technology or resources available would allow for acquisition of the required information; hence the encounter is viewed in terms of the probability of its occurrence.

To sum up, the three geographic diffusion processes are (1) purely contagious where distance or adjacency is the absolute controlling factor, (2) purely hierarchical where size or urban position in the central place hierarchy is the absolute controlling factor, or (3) where the location of change is purely random. These are the pure or ideal forms of the diffusion process; all real diffusion processes are the result of a combination of these extremes.

Illustrations of Diffusion Phenomena

Some situations from everyday life will serve to make more clear the nature of the processes under study in this book.

Example 1: Ashfall. The distribution of ashfall from the 1980 eruption of Mount Saint Helens was a diffusion process determined by such elements as strength of the initial explosion, volume and composition of material emitted, direction of explosion, and the direction of prevailing winds at various heights in the atmosphere. Using simple diffusion models, an estimate can be made of what physical forces are consistent with generating the observed pattern of ashfall.

Example 2: Epidemic. Imagine a traveler contracts a strain of influenza. The traveler returns home before the condition is realized since influenza requires a period, say ten days, of incubation. Although it is not yet apparent to others, the traveler is an infectious agent contacting those persons regularly encountered in daily life: the teacher, the store clerk, the family. The nature of the spread of the new strain of influenza will depend upon the temporal and geographic pattern of contacts, how soon the carrier becomes restricted after the disease is discovered, and the susceptibility of persons contacted with the disease.

Example 3: Language. A characteristic of a living and vital language is that new phrases are regularly coined to express new things, ideas, concepts, and feelings. Occasionally, a phrase catches on among one's peers. Its use may spread slowly at first; the phrase is initially considered as slang. It becomes used at work or school; if it is used on TV or radio, the rate of spread in the local area becomes more rapid. Its spread into other regions depends upon the number and kinds of contacts between the origin region and potential destination regions: letters, telephone conservations, and traveling. Whether it is accepted in the other regions depends upon the appropriateness of the phrase to people in other environments. This is a category in which selective diffusion may occur, that is, that the new phrase is not necessarily used by the entire population, but often only by subgroups, such as athletes, teenagers, or musicians.

Example 4: Microcomputer purchase. Adoption of a microcomputer depends upon contacts people have with others that may have already adopted a microcomputer or used them in work or school, although the efforts of advertising and salesmen may influence purchasers. Along with other elements, some considerations of the adopter may include: the cost of the microcomputer, disposable income after other necessary expenditures have been made, the potential adopter's general predisposition toward microcomputers and high technology, and the effectiveness of advertising.

Example 5: Franchises. Consider a retail outlet that has developed a unique product identity; the uniqueness can be attributable to, for example, a new manufacturing process or design. An entrepreneur considers that additional outlets could be added. The rate at which new outlets are added is restricted by capital availability and the market. The capital problem is overcome by using other investors' money by selling franchises: the right to duplicate the process of manufacture or the right to distribute the good and use the name of the product or store. To add to already existing demand, and to increase the desirability of using the name of the product, the entrepreneur spends a large amount of the proceeds on advertising. The diffusion is carefully planned, considering capital, size of additional markets in relation to competition already there, and access to and ease of administering more distant outlets. It is interesting to note that the locational goals of the franchiser and the franchisee are opposite: Since the franchiser usually earns a percentage of franchise sales, he wishes to saturate the market; conversely, the franchisee seeks to carve out as large a competition-free market area as possible. Examples of such franchises include restaurants, most notably fast-food hamburger stands; computer stores; ice cream stores; health clubs; and even hairdressers.

Example 6: Expansion of a ghetto. An ethnic enclave within a city, if the population grows enough, must expand into existing urban settlement. This process differs from those above in the likelihood of strong formal and informal resistance to the expansion process from government, neighborhood and ethnic organizations, as well as individuals (Morrill, 1965).

Example 7: City growth. Cities expand both vertically and at the periphery of already built areas. At the urban margins, land where use is largely nonurban, perhaps agricultural, and sparsely settled, becomes subdivided. Roads, utilities, and other urban infrastructure are added. The rate of growth depends upon the availability of population to relocate to the newly developed area, the availability of capital, the ability to zone the land for development, owner and developer evaluation of potential profit, the entrepreneur's information about opportunities, and the actions and views of friends and associates as well as the general value system of society.

Note that examples 1 through 7 are listed in order as a gradual progression from dependence on the individual's knowledge of and interaction with those around them, or chance occurrence within the physical environment, to instead a greater dependence upon economic, legal, institutional, and general societal constraints. Still, in every example, the surrounding environment and patterns of personal and commercial contacts are prerequisites to the human geographic diffusion process.

What is the common thread? Simply the idea that the movement of many phenomena represents a diffusion or spread of characteristics from limited origins into a wider territory. The unfolding process is governed by the existing spatial patterns of people, information, opportunity, or capacity for change. In turn, the process of diffusion may change the landscape over which it has occurred, if only briefly.

In the next chapter, nongeographic approaches to diffusion will be reviewed. Then the basic diffusion model that includes geography—the spatial dimension—will be presented. The fourth chapter will present refinements on the basic model and examples and the fifth chapter will focus on mathematical approaches to spatial-temporal diffusion models. The book will conclude with an examination of the present status and future needs of spatial diffusion research.

2. NONGEOGRAPHIC DIFFUSION LITERATURE

Since diffusion phenomena are so pervasive, they are a necessary subject in many fields. This chapter will review the efforts in several academic disciplines toward the explicit incorporation of diffusion phenomena into their subject of study. It should be noted that many of these studies are temporal, thus discounting the importance of the spatial dimension in the study of diffusion. Nonetheless, the robustness of the diffusion paradigm can be inferred from the wide variety of its academic applications.

Physical Sciences and Mathematics

An early scientific treatment of diffusion phenomena was the analysis of the movement of gasses between two vessels; at a larger scale, atmospheric scientists are concerned with the horizontal and vertical movement of large parcels of air. Most of the physical sciences have at one time placed great effort in the development of their particular applications of the general diffusion process. It is largely from these fields that the mathematical models of diffusion have been developed that in turn have proven useful in human-environment studies.

Mathematicians and statisticians have treated diffusion processes as abstract theoretical problems (see Stroock & Varadhan, 1979; Yuill, 1964). Recent work includes Karatzas's (1984) study of a theoretical dynamic allocation problem, McKeague's (1984) estimation of parameters of diffusion under misspecified models, and Shreve, Lecaczky and Gaver's (1984) study of optimality with respect to barriers. Operations researchers (such as Filipiak, 1983; Kimura & Ohsone, 1984) have incorporate diffusion models into queuing theory. Lamm (1984) literally adds a new dimension to the study of diffusion in his analysis of three-dimensional diffusion in Brownian dynamics.

Biological Sciences

Biologists are concerned with the spatial distribution and differentiation of species of animals and plants, including the process and effect of the introduction of new species into the environment; animal and plant reactions to "catastrophic" changes in the environment; the exploration behavior of animals in the search for food; and the spread of animal and plant diseases. Particularly relevant to the social sciences has been the mathematical work of the biologist Pielou (1969) concerning the mathematics of spatial processes. This work has proven valuable to the study of human exploration, settlement relocation, and the dispersal of plants and people (see also McMaster, 1962; Sauer, 1952).

Neurophysiologists have used temporal diffusion models to analyze spontaneous neuronal activity (see Lansky, 1983; Ricciardi, 1977).

Epidemiology, the study of the spread of disease through populations, can be considered a subfield of biology. In epidemiology theory there has been some concern with spatial constraints or channeling of epidemics. In particular, Bailey (1968), Kendall (1965), Bartlett (1975), and Mollison (1977) have modeled the role of population density, population migration, and the necessary critical mass of population required for an epidemic to occur. An important aspect of the epidemiological tradition is that of the death of the diffusing agent or recipient; this death mechanism in turn affects the rate of further spread and the termination of the epidemic. The literature dealing with the theory of epidemiology has generally proceeded without introducing the component of space.

In contrast, field epidemiological studies investigate specific epidemics on a case by case basis; the location of the victim is crucial in these studies. Adequate geographical data has generally not been collected on epidemics; there are few instances of attempts by researchers to construct a geographic data base of an exact person-to-person contact chain of contagion; one of the more successful attempts has been done by Angulo (see Morrill & Angulo, 1981) of a smallpox epidemic in Brazil.

Sociology

Diffusion has been a minor but continuing tradition in sociology, including analyses of social networks (Dodd, 1950) and the neighborhood effect (McVoy, 1940), the spread of rumors (Dodd, 1950), the emergence of opinion leaders, and the nature of social resistance, including attitudes toward birth control. This literature, in turn, has influenced subsequent studies in agricultural economics including the classic study by Griliches (1957) on the diffusion of hybrid corn in rural sociology, and Hagerstrand's (1953) monumental work in agricultural geography, which will be reviewed in depth in later chapters.

The work by the communication sociologist Everett Rogers (1962, 1971, 1983) has emphasized the role of information, communication, formal and informal media, opinion leaders and social networks, and economic and psychological constraints on acceptance. Rogers's work stresses the decision mechanism of the potential adopter; the work of Rogers partitions the adopter's process of choosing to accept a new phenomena or trait into five stages: stage 1—the potential adopter gains knowledge or awareness of an innovation; stage 2—persuasion is exercised to adopt; stage 3—a decision is made to adopt; stage 4—the decision to adopt is implemented; stage 5—the adopter confirms the decision to adopt. In Rogers's stage theory there may be a substantial time lag between when a potential adopter becomes aware of the new characteristic and when a decision is actually made to adopt. In empirical studies Rogers found that for the same phenomena, innovators accepted the new characteristic in an average of only five months while laggards required years before adoption took place. To Rogers, communication patterns through friendship networks are the predominant avenue of innovation diffusion.

Sociologists are also concerned about the consequences of the diffusion of some new phenomena upon society. The effects can be both positive and negative. Rogers (1983) traces one possible scenario of wet rice cultivation in Madagascar. One trajectory is for changes in labor techniques to be instrumental in creating the breakdown of kinship clans; lineages take on only ceremonial importance; bonds are merely for economic gain; many households outmigrate.

Economics

To the economist, a major theme of study is the role of the innovator and the innovation; how is the innovation created; what kinds of persons become innovators; what kinds of places or environments are conducive to innovation; how do technological innovations become diffused among and within organizations, including governments and business firms. For a review of this literature based on the seminal work of Schumpeter (1934), see Davies (1979).

Diffusion plays a major role in the study of economic development. Advocates of neoclassical models of economic growth argue that technical change induces growth and that the diffusion of technical change will cause a broader macroeconomic growth and thus reduce inequalities (Metcalfe, 1981; Soete & Turner, 1984). This approach is similar to that of the modernization school in geography (Berry, 1972; Gould, 1970; Soja, 1968). Forbes (1984) provides a fine critique of the diffusionist approach to development.

History

The historian too has a tradition of diffusion themes in the literature. Webb (1927) believed there to be a frontier character and mentality that drove the American westward, spreading their culture and settlements across the central plains. In Europe a parallel to the frontier mentality is the rise of the factory discipline and the Protestant mentality, social forces that drove the innovator to new creations (for a review see Landes, 1969). The effect that the historical school has had upon geography has not been insignificant; the historical theme was developed in geography by Isaiah Bowman (1931) in his geographic analyses of pioneer settlement (advance and retreat) in many parts of the world.

Anthropology

Anthropology has had the greatest outside influence upon the development of the diffusion literature in geography. Central to anthropology is the study of the evolution of the human landscape, particularly the spread of humanity, language, religion, technology, agriculture, and urbanization (for a discussion see Edmonson, 1961). Two broad theories that have emerged from the anthropology literature on human social evolution are relevant here: spontaneous evolution in isolation of external interaction, namely, evolution *in situ*, and evolution by diffusion, that is, did agriculture arise independently in many places, or did it develop at one hearth and then diffuse. Precise measures for the diffusion of language and bloodtype have been used to support the diffusionist school. However, the diffusion of pottery, agriculture, and urbanization have not been so successfully measured; the weight of evidence is not conclusive toward either school of thought.

The great anthropology debate has been one of the stimuli toward development of cultural geography; the most well known example of this type in geography is the work by Sauer (1952; Sauer & Brand, 1930).

Sauer proposed that hearths of plant domestication can be identified on the basis of natural and cultural conditions; migrating people have taken their plants with them. Hence historical migration routes can be traced by analyzing the spread of domesticated plant types back to their hearths or origins. The origins of vegetative reproducing plants, seed plants (new and old world), and animal husbandry have been compiled by Jordan and Rowntree (1981); their work is based in part upon the seminal work of Sauer.

Conclusion

The previous discussion raises three questions, critical to the study of scientific geography in general, and specifically spatial diffusion: (1) How does the differential character of space or place generate phenomena that may in turn spread? (2) How does spatial separation and spatial structure of the landscape influence the subsequent spread? (3) What is the balance between the mechanisms of diffusion on the one hand and the uniqueness of local places on the other to driving the evolution of the landscape?

There is a short answer to these three questions. Diffusion is the "equilibrating force" that reduces differences between places and thereby promotes a more widespread presence of phenomena. The differences are reduced until the conserving force for the retention of the differences is equal to the driving force for their elimination. Of course, local uniqueness might also come from local innovation, that is, it may not be due to conserved tradition, but to generated innovation. In some cases local innovations are being generated more rapidly than they can diffuse and therefore locational differences are increasing ("culture" in Los Angeles may be an example of this).

In human terms, we may view the reduction in local uniqueness with some regret, such as the disappearance of a dialect or custom; however, change has always been with us, and those characteristics that we identify with today are themselves the product of centuries of spatial diffusion.

3. DEVELOPMENT OF THE BASIC CONCEPT OF DIFFUSION AS A SPATIAL PROCESS

Pragmatically, the construction of a scientific theory often begins by listing those elements that delineate the phenomenon under consideration from other phenomena. These delineators are at times assumptions that serve to focus our attention upon a small component of an otherwise large and complex reality. In the study of spatial diffusion, it is important that we have information about the following twelve elements in order to delineate the diffusion phenomenon from other phenomena in the social sciences:

(1) Is the phenomenon or characteristic transferable or adoptable?
(2) Did the phenomenon arise at one or more locations?
(3) What is the location and status of adoption by the population? Status of adoption includes (a) those who possess the characteristic (adopters), (b) those who don't possess the characteristic but could (susceptibles), and (c) those who are not susceptible.
(4) What is the means of communication or contact between either individuals or groups of individuals?
(5) What is the spatial distribution of adopters and susceptibles? Also important, and interdependent with this element, is information on the likelihood of each person being contacted by the other; this rises or falls depending upon the cost or difficulty in making contacts.
(6) What is the level and variability of adopter enthusiasm, the persistence of adopters contacting others, and the resistance (psychological or otherwise) of the susceptible population?
(7) What are the minimum economic and technical barriers for acquiring the characteristic?
(8) What is the relative attractiveness and life expectancy of the characteristic—from a short-lived fad or epidemic to a permanent change?
(9) What is the degree to which the phenomenon is appropriate to the susceptible population?
(10) Is the process better described by a deterministic or stochastic structure—how much uncertainty surrounds the decision?
(11) Are there formal "propagators" or "promoters" of the phenomenon?
(12) Are there competing phenomena?

The importance of the twelve elements can be illustrated by two examples; the first example is the diffusion of a belief system, and the second example is the diffusion of a commodity. In these examples, the relevant element numbers will be in brackets.

First, consider a religion such as the faith of the Mormons. The faith is adoptable and had a specific original location [1, 2]. The population may be partitioned into a specific distribution of those who are and who are not Mormon; the population of non-Mormons can be further classified as to their relative susceptibility as converts [3]. The likelihood of contact is affected by social networks based upon neighborhood, work, or school acquaintances [5]. Mormons vary in their evangelism and non-Mormons vary in their resistance to conversion [6]. Information about the religion can be communicated by television, radio, newspaper, and personal contact through missionaries [4, 11]. Although the religion may be more attractive to some persons than others depending upon their economic status and knowledge of technology, neither technology nor financial status are barriers in this case [7, 9]. The phenomenon can be long lasting [8]. The process contains features that are deterministic: persons whom missionaries contact may adopt given that these persons are susceptible people [9, 10, 11]. The process also contains features that are stochastic: What is the probability that an individual contacted is susceptible [10]? At a broad scale, the pattern of diffusion might be expected to, and did, spread outward concentrically from the Utah core, but it was conditioned by the nature of the economy (spread was most successful where irrigation agriculture was expanding); the strength and persistence of existing religious affiliations [12]; and the pattern of settlement (skipping quickly to large metropolitan concentrations such as Los Angeles or Seattle) (Meinig, 1965).

Second, consider the diffusion of the willingness to purchase air-conditioning systems (after they've been invented and are available at numerous locations) [1, 2]. The population may be divided into those who own and those who do not own air conditioners [3]; there are also air-conditioner dealers and advertisers [11]. Contacts may be interpersonal, as well as through television, radio, and newspaper; contacts may also be between dealers and individuals [4]. There is a particular spatial pattern of likely contacts among these groups [5]. Individuals differ as to their resistance to purchase; existing owners and dealers vary in their aggressiveness to persuade [6]. In this case, there are economic and technical constraints to adoption [7]. Also, the lifetime due to model and technological changes is of short duration [8]. The appropriateness of adopting the phenomenon may depend upon characteristics of location such as warm versus cool climates [9]. At the scale of the individual or neighborhood, there is considerable uncertainty about the likelihood of purchase [10]. Promoters may devote varying amounts of resources in convincing persons to adopt, depending upon the location and economic status of the potential consumer [11]. Finally, there are competing goods or services that could be purchased [12]. At a broad national scale, in the United States, for example, the pattern of spatial diffusion of air-conditioner use might be expected to have moved generally from south to north, and also to be affected by microclimatic variations, and even availability of electricity.

The above two examples illustrate the twelve basic elements that characterize diffusion phenomena. These elements are an expansion of the basic model derived by Hägerstrand and presented below.

Few characteristics are universal; those that are not universal can be evaluated in terms of their potential of being adopted where they currently are not adopted. A phenomenon can diffuse to regions where it currently has no hold and thereby displace competitive phenomena dominant at the present. Thus a theory of spatial diffusion should say something about (1) the origin of the phenomena, how it came to exist; (2) how the phenomena became diffused or dispersed—the nature of information, communication, and propagation; and (3) the spatial patterns of the phenomenon resulting from diffusion. The study of condition (1) concerning the origin of the phenomenon overlaps with location theory (Webber, 1984); an understanding of diffusion is interdependent with an understanding of location.

Introduction to the Hägerstrand Model

The seminal theory of spatial diffusion was set forth by Torsten Hägerstrand in his 1953 dissertation (see 1967), *Innovation Diffusion as a Spatial Process*. The importance of Hägerstrand's thesis as a contribution was not recognized in the United States until 1959-1960 when he taught at the University of Washington. Later that year, he presented his paper "Monte Carlo Simulation of Diffusion" at the National Science Foundation Conference on Quantitative Methods held at Northwestern University.

Hägerstrand considered diffusion to be a fundamental geographic process: Whatever the phenomenon being diffused might be, one may consider it in the context of a larger universal process of spatial diffusion. For something to be gradually diffused over space and time, Hägerstrand deduced that there must be a mechanism of contact and persuasion to transmit the phenomenon. People in Hägerstrand's work were the primary agents of contact; therefore, the spread of diffusion of a phenomenon reflects the pattern of contacts or acquaintances of those people.

Hägerstrand believed the contact network of most people to be quite localized; thus the diffusion of innovation should likewise be local, spreading from person to person in a contagious manner. Hägerstrand argued that since the contact network of a population remains relatively stable through time, it can be described; subsequently, with knowledge of the spatial pattern of the contact network, one can predict the spatial diffusion of most innovations.

Hägerstrand believed that knowledge about the diffusion phenomena is gained by information from the media or from the generally more pervasive interpersonal mode of communication. People vary in their resistance to adopting the diffusing phenomena. This resistance may be attributable to psychological barriers: Some people are risk takers whereas others are followers. Resistance includes economic barriers: Some can't afford it. And for some, the innovation has very little usefulness. The diffusion pattern unfolding over space and time then depends upon the spatial distribution of "information" or acquaintances of both knowers and nonknowers.

Hägerstrand discovered that most individuals have a spatially biased "field of information," rich in information about areas proximate to home, poor in information about things relatively remote from home. This spatial bias leads to phenomena being spread gradually over space, contagiously from one person to the next. Hägerstrand referred to this bias toward local information as the "neighborhood" effect. This conception can be visualized as a pebble thrown into a pond resulting in waves radiating out from the core or origin; a wave or frontier of more active acceptance of change spreads outward from the origin of the diffusion phenomena. Eventually the population of susceptible persons becomes "saturated" when no susceptible persons remain that have not adopted the diffused phenomena. This spatial component of diffusion is considered Hägerstrand's most significant specific geographic contribution.

Hägerstrand also formalized diffusion as a temporal process. In his temporal model, the diffusion process is partitioned into periods: an early period of pioneering, a middle period of diffusion and fastest change, and a later period of "condensing" and saturation, or filling in. In turn, the temporal periods can be mapped onto the "S-shaped" or logistic pattern of the cumulative proportion of persons having accepted the diffusion phenomena over time (see Figure 1).

Upon reflection, Hägerstrand's theory appears not only intuitive but obvious: (1) phenomena spread across a territory or population mirroring the spatial pattern of relations among people; (2) there is a match between the phenomenon being diffused and peoples' needs; and (3) diffusion in any locale begins slowly at first, then rapidly spreads through the population, then becomes slow again when the majority of persons have adopted the phenomena. That it was intuitive and obvious once it was clearly stated underscores the greatness of Hägerstrand's contribution: No one had clearly expressed the idea before as a general principle that could be applied to diverse phenomena, gathered such thorough and convincing evidence, and demonstrated that it was possible to model and reproduce historical processes.

To achieve a full appreciation of Hägerstrand's contribution, the reader is urged to read his book. While some consider Christaller's book on central place theory (for a review see King, 1984) to be the greatest contribution in geography in this century, others give this accolade to the book by Hägerstrand. Probably both deserve to share this recognition.

Why? First, in 1953 when Hägerstrand's book was written, geography was essentially a descriptive subject, as opposed to being predictive and prescriptive. Second, Hägerstrand's analysis is, on the one hand, a deductive and creative theoretical leap: how a process should occur, based on simple notions of space and behavior. On the other hand, the work represents a thorough amassing of empirical data on actual cases of diffusion that were used to verify the theoretical structure. A third contribution that was not limited to spatial diffusion was his pioneering use of Monte Carlo simulation that permitted the modeling of individual behavior to create collective patterns.

Theory and Models

Hägerstrand created several models to operationalize the theory of spatial diffusion. Model I assumes that all persons are aware of an innovation and that acceptance occurs independently of one another in random order; consequently the pattern of adoption over space is random. Model II introduces the "mean information field" (MIF); acceptance occurs immediately upon contact from an earlier adopter. Model III incorporates resistance barriers; it is the most complete and will therefore be discussed here. The reader is referred to Moryadas (1975) for a more thorough and advanced treatment.

Given a limited distribution of initial adopters, and a distribution of population (potential adopters), how would diffusion take place? Hägerstrand proposed the following five assumptions or "rules of the game":

First, partition the diffusion process into discrete time periods, such as years, and the region into regular "cells" or other subdivisions.

Second, in each time period adopters are required to make a finite number of contacts with other persons; this may be as small as one or quite large.

Third, the location of the contacts reflects the acquaintance fields around each of these initial adopters. Since individual acquaintance fields are not precisely known, an average field referred to as the "mean information field" (MIF) is used to describe the probability of contact at different distances and directions.

Fourth, this MIF may be affected by physical barriers, such as lakes or mountains.

Fifth, potential adopters must receive a certain known number of contacts before they adopt the diffusion phenomenon; this may be as small as one or several may be required. The process is repeated for the next time period; in the subsequent time periods, those who adopted in the earlier time period also make contacts within their contact fields. This process continues for several time periods. Early contacts are critical in establishing corridors or directions of spread. Eventually areas with larger numbers of possible adopters will tend to have more adopters and contactors.

Each knower (innovation adopter) is surrounded by many potential contacts. A general MIF for people in the study area is constructed; it summarizes the probability of a contact in a particular direction about a person who has already adopted the phenomenon. The MIF is determined in this case by the historically observed pattern of local migration; the number of migrants diminish with distance from the origin. To illustrate the creation of the MIF, Hägerstrand's original effort will be recreated in a step-by-step fashion. The problem is the simulation of the migration of 248 people in the Asby area of Sweden.

Step 1: A 5 × 5 square grid is used to "map" the study area. Each grid cell is 5 kilometers square.

Step 2: Grid squares are identified by coordinates. From the top left corner, numbers 1 through 5 identify the vertical axis from the top down, and a through e identify the horizontal axis from left to right. Grid cell 3c is the center grid cell and is always the origin.

Step 3: The distance, D_j, from the center of each grid cell to the center of the origin cell (3c) is measured.

Step 4: Migration is calculated to diminish with distance by the gravity formula:

$$M_j = 0.76 \, D_j^{-1.59} \qquad [3.1]$$

where M_j is the number of migrants to any grid cell j at distance D_j; 0.76 is a constant based on the total number of migrants that translates between number of people migrating and units of distance;

and −1.59 is an estimate for the slope or rate of distance decay. Note that the constant and exponent vary depending on the case (the reader is referred to Haynes & Fotheringham, 1984, for a discussion of gravity models).

Step 5: The distances calculated in step 3 are inserted into the formula in step 4, allowing a calculation of M_j, the number of migrants expected in each grid cell.

Step 6: The expected numbers of migrants, M_j, are entered into their appropriate grid cells in Figure 3.1 Note that the values in Figure 3.1 are symmetric, given that the distances of the grid centers from the origin are also symmetric. According to this formulation, 110 of the 248 migrants would migrate to new locations within the center grid cell, while only 2.38 would be expected to migrate to a specific corner grid cell. The sum of migrants for all 25 cells should total to 248.

Step 7: The value of the expected number of migrants, M_j, for each cell in Figure 3.1 is divided by the total number of migrants (248). This proportion is inserted into a new grid in Figure 3.2. This new grid is the famous "Mean Information Field" (MIF).

The proportions or probabilities in this grid should sum to 1.0. This means that all migration occurs within the region and is accounted for. The MIF in Figure 3.2 can be interpreted as predicting that a little over 44% of migrants can be expected to migrate within the central grid cell (3c), while less than 1% of the migrants can be expected to migrate from the central cell to one of the corner grid cells (for example, 5e).

The MIF created in Figure 3.2 can be used to simulate the diffusion process of the migrants. Simulation is used in order to consider which *patterns* of migration might be expected. If the migration process is deterministic, only *one pattern* of migration would be expected. Herein the migration pattern is stochastic (has a random element) and many *patterns* are possible. The stochastic element in this migration example is based on human choice behavior. A migrant may choose to move to a new loction within the central grid cell or to move to another new location in any other grid cell. The model argues that distance from the origin predicts the likelihood of any of the migrant's choices.

The step-by-step example will be continued to illustrate a simulation using the MIF in Figure 3.2. The simulation uses a "Monte Carlo" technique.

Step 8: Choose a decimal base (100, 1,000, 10,000, and so on) larger than the number of cases (that is, > 248 here). In this example a 10,000 base is used.

	a	b	c	d	e
1	2.38	3.48	4.17	3.48	2.38
2	3.48	7.48	13.57	7.48	3.48
3	4.17	13.57	110.00	13.57	4.17
4	3.48	7.48	13.57	7.48	3.48
5	2.38	3.48	4.17	3.48	2.38

SOURCE: Lowe and Moryadas, 1975.

Figure 3.1 Observed Local Migration in the Asby Area, Sweden; Standardized by Symmetrical Cells

	a	b	c	d	e
1	0.0096	0.0140	0.0168	0.0140	0.0096
2	0.0140	0.0301	0.0547	0.0301	0.0140
3	0.0168	0.0547	0.4431	0.0547	0.0168
4	0.0140	0.0301	0.0547	0.0301	0.0140
5	0.0096	0.0140	0.0168	0.0140	0.0096

$\Sigma = 0.9999$

SOURCE: Lowe and Moryadas, 1975.

Figure 3.2 Mean Information Field, Asby Area, Sweden

Step 9: Multiply the probability of the first cell (1a) of the MIF by the decimal base. In this case, .0096 × 10,000 = 96.

Step 10: Assign sequential integers to cell 1a equal to the number calculated in step 9. Thus integers from 0001 to 0096 are assigned to cell 1a.

Step 11: Repeat step 9 for every cell. Repeat step 10 also, but assign the integers to follow sequentially those assigned to the previous cell. Thus for cell 1b 0.140 × 10,000 = 140 integers are needed. These 140 integers, however, must begin with 0097, since 0001 through 0096 have already been assigned to cell 1a. Thus the integers assigned to cell 1b are 0097 through 0236. Figure 3.3 illustrates the completed grid.

Step 12: Find a random number table in a statistics book or a book of mathematical tables.

Step 13: Select a random point to begin in the table (you can use sophisticated methods or close your eyes and pick a starting point with a pencil).

Step 14: Go through the table selecting one random number for every case. In this example 248 4-digit random numbers are selected.

Step 15: Assign the new locations for every case. Since the random number selected for the first case is 9,292, this means the first migrant's simulated move is from the origin cell 3c to a new location in cell 4e since the random number falls in the range (9,220-9,359) for that cell. The second migrant's random number 8,220 locates his move in cell 4b.

Step 16: Construct a map of the simulation by counting the number of cases (migrants) located in each cell.

This is a basic simulation for one time period. In the next example, the model will be refined to include subsequent time periods. Barriers and resistance will be considered in the next chapter. The model is simple and at the same time consistent with the theoretical arguments.

Hägerstrand next proceeded to an empirical analysis and then applied the model to the data in order to simulate the actual process. The simulation is for more than one time period. While the migration example above was for one time period, many diffusions occur over several time periods. Finally Hägerstrand tested the correspondence of actual and simulated patterns.

The Empirical Studies

Hägerstrand investigated three agricultural innovations: (1) the innoculation of cattle, (2) improved pasture subsidies, and (3) soil mapping. The pasture subsidy maps will be illustrated here.

The actual pattern of the acceptance of pasture subsidies in a section of southern Sweden is depicted in Figure 3.4 The process shows a tendency for local neighborhood spread over time; also new centers of spread are generated by the occasional long distance leap. The temporal analysis also revealed the by now familiar S-shaped curve. The process began in 1928, developed slowly until 1930, and spread rapidly in 1931

	a	b	c	d	e
1	0001 .0096	0097 .0236	0237 .0404	0405 .0544	0545 .0640
2	0641 -.0780	0781 -.1031	1082 -.1628	1629 -.1929	1930 -.2069
3	2070 -.2237	2238 -.2784	2785 -.7215	7216 -.7762	7763 -.7930
4	7931 -.8070	8071 -.8371	8372 -.8918	8919 -.9219	9220 -.9359
5	9360 -.9455	9456 -.9595	9596 -.9763	9764 -.9903	9904 -.9999

SOURCE: Lowe and Moryadas, 1975.

Figure 3.3 Grid to Simulate the Role of Distance in the Simulation Model (floating MIF)

and 1932, and tapered off until 1938. Hägerstrand then used the Monte Carlo model to simulate the diffusion process.

The Monte Carlo Model

Conditions for the Model

(1) Time is divided into even, meaningful units (Hägerstrand used years in his agricultural studies) and space into simple, regular units (cells) if possible. (Spatial barriers will be discussed in Chapter 4.)
(2) Existing adopters do contact and inform others, but
(3) they make only a few contacts per time period due to constraints on their time and travel and the normal limit of the number of friends they have made that are available to contact.
(4) Their acquaintances and thus their potential contacts can be described by a distance decay function. That is, the mean information field (for example, Figure 3.2) truly expresses the probability of contact at any distance and direction.
(5) Potential adopters vary in their resistance; some may require more than one "telling"; physical barriers may also inhibit contact.
(6) Since existing adopters do not have the time to make thousands of contacts, nor would normally choose to do so, but only time and preference for a few contacts, a "Monte Carlo" sampling procedure is used to generate a set of opportunities surrounding each adopter.
(7) The process is iterative, with each period building on the result of previous periods.

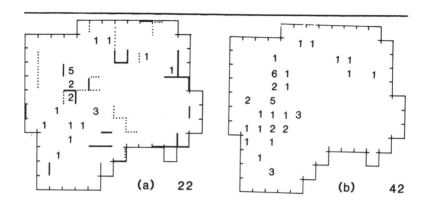

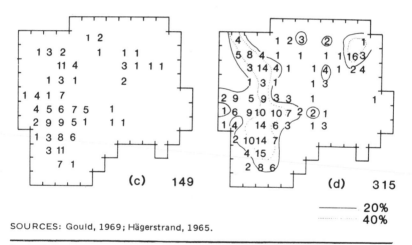

SOURCES: Gould, 1969; Hägerstrand, 1965.

Figure 3.4 Actual Diffusion of Pasture Subsidies

Monte Carlo modeling involves taking a set of randomly selected numbers to represent individual contacts. Each number corresponds to a contact at a given distance and direction from some existing adopter, and reflects the known decline in probability of contact as distance increases. Although any one individual may make only one contact, if we looked at hundreds of contacts, we would find that their distribution does reflect the frequencies of numbers in the mean information field. Each individual contact is unique, but the whole set of hundreds of contacts would show the expected proportions at each distance and direction.

This is an ingenious way to simulate a microgeographic process. The word *simulate* means to imitate a complex reality using only simplified relationships from that reality; the goal of simulation is to imitate real patterns using only small amounts of information. In this context it is not necessary to have detailed information as to who contacted whom. Hägerstrand argues that it we are to understand spatial diffusion as a general process, then it is not essential to know the detailed history of a particular case, but it is essential to find the underlying mechanism; he believed that the seven assumptions listed above describe that process.

Testing the Model

Several tests that replicate Hägerstrand's data and simulations have been done. Cliff (1968) failed to confirm a neighborhood effect on the original data (pasture grazing subsidy), but Haining (1982), using a nearest neighbor (whether a new adopter was a nearest neighbor of an earlier adopter or not) spatial interaction equation, found a fairly strong neighborhood effect, as did Morrill (1970) using both a polynomial and an exponential equation in distance and time. In Morrill's study, distance of a new adopter from an earlier adopter was a highly significant variable (pasture subsidy and inoculation for tuberculosis). Barton and Tobler (1971) used spectral analysis to reveal a wavelike character of diffusion (Hägerstrand's original data). See the mathematical analysis in Chapter 5 for further details.

The question of whether simulated patterns were sufficiently like the actual for one to conclude that the diffusion process was correctly modeled is an important but difficult problem. (Note that here the reasoning is that the real world is regarded as one logical outcome of the process (mathematical model) that produces the simulations.)

Hägerstrand's basic contribution was the idea that there is some measure of regularity or predictability, even at the microgeographic scale, in a whole class of diffusions of geographic phenomena over space. Because most examples of spatial diffusion are so complex, it is rarely possible to reproduce them exactly. Rather, Hägerstrand endeavored to discover the rules that govern the essence of diffusion, setting aside local idiosyncracies; but the fewer local details taken into account, the greater the number and variety of spatial patterns that can occur given the same rules or model. The critical question then is whether the actual diffusion, with its idiosyncracies, is consistent with the rules captured in the model.

In general, the logic of this statistical test is that a large number of simulations are run, various measures of the spatial structure (centrality, dispersion, pattern) are calculated, and it is then determined whether the structure of the actual pattern is significantly close to the simulated pattern. To corroborate that the distribution of frequencies of adoption by cell was the same for simulated and actual patterns, Hägerstrand used the chi-square and Kolmogorov-Smirnov statistical tests. More contemporary work has used the method referred to as quadrat analysis.

In quadrat analysis (Thomas, 1976) a surface is partitioned into a regular grid of cells; the frequencies of occurrences of phenomena in adjoining neighbor cells are then measured (for a discussion see Boots & Getis, 1988, in this series). The test evaluates simultaneously frequency of occurrence of the phenomenon, and spatial pattern of that phenomenon. For example, quadrat analysis reveals whether the phenomenon has tendencies toward dispersion, clustering, or randomness. Quadrat comparisons of Hägerstrand's original simulations indicate a roughly comparable degree of clustering or concentration between the observed distribution of phenomena and the simulated distribution.

Other statistical tests that have been used to compare observed with simulated diffusion patterns include measures of spatial autocorrelation. Tests of spatial autocorrelation include Geary's C (1954) and Moran's I (1948) tests (for a discussion see Odland, 1988, in this series). The tests have indicated that the spatial pattern has a higher autocorrelation (or degree to which adjoining cells have similarly high or low frequencies) than the observed diffusion phenomena; in other words, simulated diffusion phenomena show high levels of regularity while observed or real diffusion phenomena demonstrate a greater presence of randomness.

Harvey (1966) provided an early quadrat test of Hägerstrand's diffusion simulation, testing actual patterns against patterns that would be expected from diffusion processes that were suggested from theories of plant propagation and concluded that Hägerstrand's simulation exhibited too great a concentration of adoption.

The above tests are tests of the similarity of general pattern. Alternatively, one can test the "goodness of fit" or correspondence of the maps, cell by cell (Tobler, 1965). Cliff and Ord (1973) suggest creating an average frequency map of many simulations, and comparing individual simulations and the actual diffusion with the average via appropriate statistics (autocorrelation). This may be invalid, since an average map of a moderately random process is commonly meaningless, that is, unless the diffusion is virtually a deterministic one. Nevertheless, all evidence seems to support their conclusion that Hägerstrand probably incorporated too rapid a distance decay in the probability of contact and underestimated the role of propagators (Haining, 1982), and that the simulations are probably different from the actual in being too concentrated.

In testing diffusion results, rather than have an obsession with exact locations, one should test relative location and pattern. This is not easy because we tend to think in terms of the "real" map. This point is so important that it will be rephrased: The greater the degree of randomness or uncertainty, the less useful is the idea of comparing actual maps, and the more essential it is to evaluate spatial patterns. This is why nearest neighbor or quadrat tests are superior to ones of map correspondence. For further tests of simulation, see Tinline (1971), Landford (1974), and Yapa (1975).

Conclusion

Earlier in this chapter, Hägerstrand's general contribution to the development of scientific geography was noted. The significance of Hägerstrand's work to the specific study of diffusion lies, first, in his recognition that diffusion is a pervasive spatial process; second, in his demonstration that spatial diffusion exhibits a degree of regularity or predictability; third, that the actual process of diffusion in space and time critically depends on the pattern of contact among individuals, and on other sources of information available to individuals; and fourth, that it is possible to model the microgeography of diffusion as person-to-person spread. While Hägerstrand's main theoretical contribution was the very concept of diffusion as a predictable space-time process, especially his demonstration of contagious diffusion in the context of the spread of agricultural innovation in Sweden, he is probably best known to students for the actual modeling framework.

4. SPATIAL DIFFUSION PROCESSES: POST-HÄGERSTRAND DEVELOPMENTS

In this chapter the theoretical work on diffusion inspired by Hägerstrand is reviewed. Chapter 3 summarized the basic Hägerstrand model and its evaluation. This chapter proceeds initially from simple *extentions* of elaborations of the basic model (theme 1 below), through *generalizations* of the essential ideas of spatial diffusion (themes 2 through 4), to broader *implications* of diffusion (themes 5 and 6). The discussion shows how, over the years, spatial diffusion has broadened from a focus on a particular model—the Hägerstrand Monte Carlo simulation—to a major conceptual theme and modeling interest in geography. This work can be divided into six main themes:

(1) elaborations and extentions of the basic Hägerstrand conception; this includes the structure of the mean information field, the role of barriers, the nature of resistance, and measures of some characteristics of spatial diffusion
(2) exploration of alternative and broader models of diffusion, deterministic as well as stochastic
(3) introduction of hierarchical diffusion processes
(4) specification of the relation between spatial diffusion and spatial interaction
(5) recognition of the role of the propagator, markets, and infrastructure
(6) application of diffusion to city and regional development planning

Literature from these six partitions will be discussed in sequence.

Elaboration of the Hägerstrand Model

As a direct result of Hägerstrand's North American tour discussed in Chapter 3, Pitts (1963) and Marble (1967) developed computer algorithms for the model; these exercises have proven particularly useful in classrooms.

Other early work in the 1960s dealt with the nature and use of the "mean information field" (the MIF) as a generalized pattern of contacts; the MIF simulates the critical mechanism by which the actual spread of diffusion is accomplished. Marble and Nystuen (1963) showed that one could incorporate distance and directional bias in acquaintance fields that mirrored real settlement patterns such as city sectors by modifying probabilities in the MIF to reflect these biases in information. Morrill and Pitts (1967) used several samples to determine the generality of the spatial form for distance decay of information fields; they determined which mathematical representations best expressed the pattern. They demonstrated that the same acquaintance fields could be determined by many kinds of socioeconomic data; hence the contact field of persons was quite spatially restricted. Mayfield (1967) examined contact fields in India to determine if generalities about contact fields in the United States also held for less developed countries; he found that the level of economic development did not significantly alter the contact field. Brown's (1968) study, based upon the earlier work of Rapoport (1951) who analyzed the pattern of television adoption in Sweden, determined that direct personal contact fields and indirect telephone contact fields were largely the same.

Resistance

Hägerstrand's own work, and that of Rogers's concerning agricultural innovation, revealed not only that most people need persuasion, that is awareness, demonstration, and trial, before acceptance, but that individuals offer different degrees of resistance to innovations. This resistance affects the timing of acceptance, the resulting spatial pattern, and the degree to which the population eventually becomes saturated. Less resistant people may adopt the innovation well ahead of the main wave or the time of the highest rate of adoption. More resistant people may not adopt until the main wave has long since passed by.

What determines resistance? Resistance is of two forms: First, people differ as to their levels of innovativeness or willingness to be risk takers; this variability in a population is reasonably described by a normal distribution (Figure 4.1). The left tail of the distribution indicates that there are relatively few pioneers or innovators. Similarly, the right tail of the distribution indicates that there are relatively few diehard resisters or laggards.

Second, time can act as a surrogate for certain independent effects. Early on in the process of diffusion the general level of skepticism and resistance may be high. After some critical threshold level of adoption is reached a "bandwagon" effect of acceptance occurs. These two effects tend to exaggerate the unevenness of the process over time: slow at first because there are few innovators or risk takers, then explosive as the bandwagon effect takes over, and finally slow again as the population becomes saturated with adopters and increasingly scarce in susceptibles. The generalized effects of these processes yield the familiar logistic curve (see Figure 1.1) of cumulative acceptance over time, which is easily derived from Figure 4.1 in which early adopters (innovators) lead the way at the start (bottom left) of the S-shaped logistic curve and laggards adopt last (top right) of the S-shaped logistic curve. This clear delineation of process easily lends itself to more precise mathematical formalization and simulation; the operationalization of which will be done in the next chapter of this book.

Hägerstrand's (1967) more complex model divided the population into five classes: Each class was characterized by different numbers of contacts required before adoption occurred from very low (innovators)

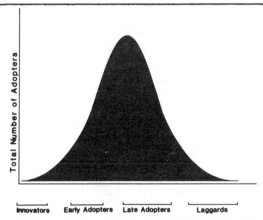

SOURCE: Figure 1-10, page 13, from THE HUMAN MOSAIC, Third Edition, by Terry G. Jordan and Lester Rowntree. Copyright © 1982 by Harper & Row, Publishers, Inc. Reprinted by permission of Harper & Row, Publishers, Inc.

Figure 4.1 Distribution of Adopters over Time

to very high resistance. Such classification will result in more of an attenuated wave-like dispersal of the characteristic. Just as epidemiologists find that resistance to infection due to preventive measures affects the spread of a disease, Morrill and Manninen (1975) found that the level and timing of resistance also affect the final level of acceptance and spatial extent of diffusion processes.

Barriers

In the basic model that uses the MIF to allocate contacts, *barriers* play an important role in introducing local geographic constraints to diffusion. Such barriers may be physical, such as rivers, lakes, mountain ranges, or deserts—for example, the great influence of the Sahara Desert in reducing contact; or they may be cultural, such as areas of different language, ethnicity, class, or income, as well as political boundaries. Barriers, Hägerstrand realized, were real impediments to interaction and these must be incorporated into the simulation. He argued that barriers had varying degrees of permeability, ranging from 100% or no barrier, to 0% or a total barrier. In his simulation in Model III, a total barrier of 0% permeability and a 50% semipermeable barrier were used, the semipermeable barrier permitting every other contact to get through. An alternative way of dealing with a 50% barrier in the model is to go immediately to a second random number table, or more simply, flip a coin. This way is much less restrictive than requiring another specific contact. In more general models, these constraints are taken into account directly in the computation of the specific probability fields around each origin (see Chapter 5). It will be helpful to go through an example of the model process in order to see how the spatial diffusion unfolds in space and time, and how it reflects these assumptions.

Figure 4.2a presents the initial situation; 22 adopters in the cells shown; solid lines indicate total barriers; dotted lines 50% barriers, which require two contacts to make one effective contact. In the initial period (a generation) each of the 22 cells makes one contact. MIF probabilities are centered over each original cell using the same step-by-step system used in the previous chapter. Note this problem is more complex, however, since the study area is slightly irregular, all the initial adopters are not in the center grid cell, and, of course, there are those barriers to consider.

Using the same step-by-step simulation system of the previous chapter, a set of 22 random numbers between 0000 and 9,999 are generated, and these are used to identify the distance and direction (location) of the contact from each origin. Suppose that the first 9 random numbers selected were 9,117, 7,560, 2,681, 2,299, 5,152, 2,777, 9,007, 7,888, and 6,656, for the initial adopters in rows 2, 3, and 4 (see Figure 4.2a). For the adopter at row 2, column 6, the 9,117 would indicate a contact in row 3, column 7 (though the random number grid is not illustrated, the process is the same as in the previous chapter's step-by-step example); for the adopter in row 2, column 7, the number 7,560 is across a 50% barrier and would not be immediately effective; for the person in row 3, column 9, the 2,681 would also not be effective (across a barrier westward); the 5 adopters in row 4, column 4 would result in contacts, first a lost contact left due to a 100% barrier; second in the same cell (#5,152), third, another lost contact west; fourth, a successful contact to row 5, column 5 (#9,007), and fifth, one in row 4, column 6 (#7,888); the adopter in row 4, col., 12 would make contact in the same cell (#6,656). Similarly, the next 13 random numbers would result in 11 successful contacts (note there are fewer barriers in the lower left areas)—7 in cells that already had an adopter, 4 in new cells (see Figure 4.2b)

Hägerstrand's basic model allows for simulation over time. In order to run simulations for subsequent time periods, the MIF is made mobile. To illustrate how this is done, continue with the step-by-step simulation from where it was left off in the previous chapter and add:

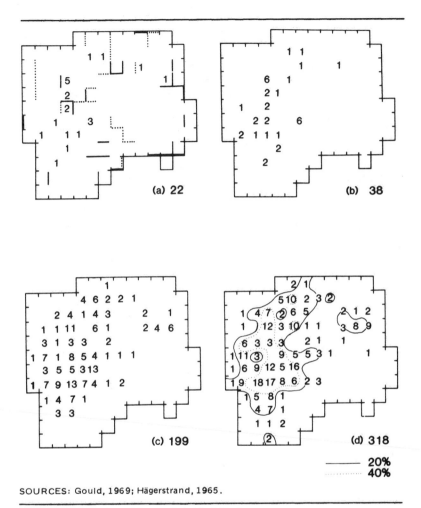

SOURCES: Gould, 1969; Hägerstrand, 1965.

Figure 4.2 Simulation of Pasture Subsidy Diffusion

Step 17: Take the MIF converted into random numbers (see Figure 3.3) and move it such that it is centered over each adopter who will contact someone during the next time period. Remember, this is a different example than in Chapter 3, but the process is the same.

Step 18: Repeat steps 13 through 16 of the previous chapter for this example.

These steps repeat the process for a "second generation," except now 38 adopters make a contact. The second- and third-generation results are not shown; they result in 30 new contacts for the second generation (now 68 total) and 47 new contacts for the third generation (now 115 total). The 115 adopters of the third generation all make contacts of which 84 are successful, resulting in the pattern of Fig. 4.2c. In the fifth generation, the 199 fourth generation adopters make contacts that result in the "final simulation" pattern of 318 adopters (Figure 4.2d). The 20% and 40% isolines indicate the proportion of the susceptible population that has adopted the innovation.

Note that the barriers are effective in reducing the influence of concentration of adopters in the NW, and in preventing heavy diffusion to the SE; the lack of barriers lead to a more rapid build up of adoption in the SW. In the main, these features hold for the actual diffusion shown in Figure 3.4 in the previous chapter. The pattern of diffusion derived from the simulation is quite similar but not identical to the actual or observed pattern of diffusion. The important theoretical and empirical question for Hägerstrand was whether the actual pattern could have been a feasible outcome of this simulation model; his challenge was to determine this feasibility in an era before computers and therefore without the ability to generate thousands of simulations.

Hägerstrand found the pace of acceptance over time to be equivalent in actuality and in simulation, as well as the frequencies of numbers of adoptions in cells; but at the time there was no statistical verification of the essential argument that there was a neighborhood effect in adoption, reflecting the fact that adoption results from spatial interaction as constrained by spatially restricted contact fields.

Barriers in Post-Hägerstrand Wave Theories

Yuill (1964) provides some intriguing examples of the effect of particular topography on spatial diffusion such as passing through a constrained opening such as a mountain pass, or around a barrier such as a lake; the timing delay upon the dispersal of the phenomenon was then measured. Other barriers were classified as reflective. For illustration, westward settlement in North America upon reaching the Pacific Coast reflected migrants in a reverse wave of settlement back into the continental interior.

Morrill (1968) introduced the theory of competing waves based on the analogy of ocean waves (Figure 4.3). The notion of "waves of diffusion" will be discussed in detail in the next section, but the barrier effects will be illustrated here. The entire impetus of the flow is reflected back in Figure 4.3a. The barrier reduces or partly absorbs the impetus

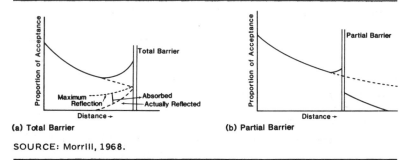

SOURCE: Morrill, 1968.

Figure 4.3 Partial and Total Barriers to Diffusion

and thereby reduces the subsequent number and success of future contacts. The proportion of the diffusion wave that is reflected after contacting the barrier actually adds to the acceptance in the area near the barrier, thus causing the proportion of overall acceptance to curve up in the area near the barrier. In Figure 4.3b, the barrier is less absolute; the USA-Canadian boundary is one example of this; cross-border flow reduced but did not stop contact or settlement from either direction.

An alternative concept of barriers can be construed from Freeman's (1985) recent work on preemption rents. In this concept the barrier to diffusion is the early adopters themselves. Freeman argues that early adopters are in the position to gain large but temporary windfall profits and may act as a barrier if it is possible for them to prevent further diffusion through political action or other oligopolistic institutional activity.

Variable Mean Information Fields

An important alteration of the Hägerstrand model was the abandonment of the unvarying mean information field (MIF) (see Chapter 3). The limited grid MIF was a precomputer device to make simulation feasible. Rather than have a constant MIF, computers can handle large quantities of information required to calculate actual probability fields around each individual location. These probability fields may change over time as well as differ for every location depending upon the number of remaining susceptibles and other characteristics. Also, in contrast with the grid cell limitation of the MIF, probability fields can be calculated for any areal unit, including counties or census tracts. Probability fields can be easily calculated by whatever formulations best express the actual contacts (see Chapter 5).

Figure 4.4 represents a polar mean information field. This modification acknowledges a critical point about diffusion—that it moves outward, usually from a center. A polar MIF allows directional influences to be more accurately and straightforwardly incorporated into the analysis through the use of a polar coordinate (versus rectangular grid) geocoding system. It is also useful in specific examples, such as urban growth outwards from the city, where a directional bias is explicit in the problem.

It is interesting to note that because Hägerstrand used a gravity model to create the probabilities of the MIF much subsequent research became more intent on the mechanics of the derivation of the MIF rather than seeing what Hägerstrand considered to be important: the idea that diffusion reflects individuals' fields of contacts and that such fields did not have to be simple and symmetrical.

Measures of Some Characteristics of Spatial Diffusion

We noted a few tests of the "neighborhood effect," of the tendency for information or contact fields around initial adopters to contain the spread, using Hägerstrand's original data. In general, an important development in spatial diffusion theory and practice was to escape from, or rise above, a preoccupation with the micro or individual behavior and exact pattern to investigate the broad idea of diffusion as a space-time phenomenon, or wave, although this broader view was itself proposed by Hägerstrand in his 1952 essay, "Propagation of Innovation Waves." This became an important theme after 1968, especially among British geographers. Blumenfeld (1954) provided an early suggestion of a wave formulation of the impact of rapid urbanization: the idea that a crest of most active adoption would proceed outward from points of origin, as a circular wave spreads from dropping a rock in water. It is reasonable to look for general patterns or shapes of diffusion.

Some of the mathematical formulations for wave spread (from wave theory and epidemiology) were summarized by Morrill (1968), who also suggested graphics of what the wave forms might look like. Consider a diffusion process over time at various distances from an origin. A wave of adoption, slow, then quicker, then slow again, approaches and passes over, but at later times, the farther from the origin. Also, if one looks at diffusion over space at various periods of time, the waves of change occupy different territories, close in at first, farther out at later periods (Figure 4.5).

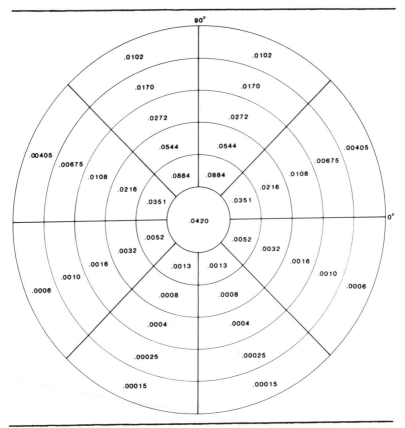

Figure 4.4 Polar Coordinate Mean Information Field with Directional Bias Such that the Directional Mean Equals 90° and the Circular Standard Deviation Equals 45°

For example, imagine that you live in a large city suburb—an area that had been rural farmland in 1910. You can perhaps visualize, starting say in the 1920s or 1930s, a slow spread of "country estates," whose owners commuted to the city. In the later 1940s, the city's built up edge rapidly approached these "country estates," and in the early 1950s the farm on which you now live was subdivided; while the zone of active subdivision is now five miles farther out beyond you, skipped-over parcels are still being built on, completing the wave of urban transformation. Or think of what was happening across the whole area, but for say just the late 1940s; perhaps in the older core of the city, little building was occurring; the zone of most intense subdivision was right at the then

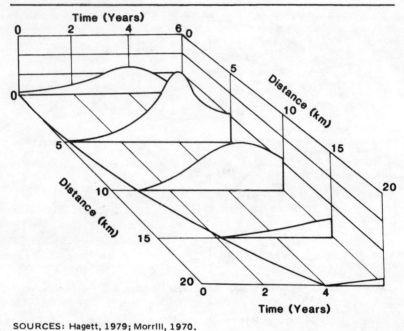

SOURCES: Hagett, 1979; Morrill, 1970.

Figure 4.5 Diffusion Waves in Time and Space

city edge, with the suburbs growing fast. Farther out, in the country, some small farms were selling out, a few becoming small subdivisions.

These ideas can be combined in a space-time graphic (here Figure 4.5, borrowed from Haggett, 1979) of the spread of diffusion from an origin as a moving wave of changing shape. Figure 4.5 is in fact a three-dimensional depiction of such a trend surface of the spread of acceptance of grazing pasture subsidies (Hägerstrand's original data). This may also be depicted as a trend surface where the two dimensions are distanced from the origin and time (Morrill, 1970).

To the degree that a wave conception is meaningful, the following characteristics are suggested:

(1) The spread of diffusion moves forward like a wave, because the initial adopters necessarily make contacts outwards from themselves, and in turn those contacted very likely have contacts yet farther away. But there is no sharp edge of new contact, because people may vary in their resistance or susceptibility, and because the earlier, close-in adopters continue to make contacts.

(2) An S-shaped curve and pattern of acceptance over time exist, not only overall, but at each distance from the origin, at later periods of time. A good example of this is shown in Figure 4.5.

(3) A distance decay is observed in the ultimate level of saturation from point of origin; this indicates a reduction in appropriateness; a weakening of the force of the innovation (as with disease or obsolescence); or confronting a competing institution—for example, language or religion. While this may be the more common situation, there are also many phenomena, like television in the United States where saturation becomes virtually total.

Many empirical tests have been undertaken. For example, Morrill (1970) tested the existence and shape of waves for Hägerstrand's original data for pasture grazing subsidies, tuberculosis controls, and for ghetto spread. The tests show curves that are from fitted regression equations (see Chapter 5) in time and distance (polynomial trend surface in distance and time, and Poisson-like equations, from wave theory), which strongly support a wave-like behavior.

For example, for both tuberculin control and pasture subsidy, the distance at which the greatest amount of acceptance occurred does shift outward, but there is much change in front of and behind this most active zone. (Unfortunately, data are for only the early part of the tuberculin-control diffusion.)

Casetti and Semple (1969) discerned a space-time wave of diffusion of tractors in the Midwest, through a generalization of the logistic equation, which included distance. Cliff and Ord (1975), using location but not distance from a supposed origin (North Dakota), found distance to be a weak variable.

Other important wave studies include those of Cliff, Haggett, and Graham (1983), Gilg (1973), and Cox and Demko (1968). Cliff et al. (1983) found evidence for waves in Icelandic epidemic data. Cox and Demko (1968) discovered that there were two separate waves of agricultural unrest in Russia, one from the southern Ukraine, and one from the Baltic coast—a case of peripheral unrest heading toward the core.

Stochastic and Deterministic Conceptions of Diffusion

Hägerstrand's conception that spatial diffusion was in part a stochastic process, involving a degree of uncertainty, was truly revolutionary in geography, though it had been incorporated earlier by way of the

Monte Carlo model in biology and zoology. The point made by Hägerstrand was not that there was not an explanation for each decision. Rather, Hägerstrand's objective was at a more general scale; the decision to contact specific individuals appears random from the vantage of an outside observer. The random element represents that part of individual behavior not known to the modeler. At the same time, the stochastic process should not be considered a "stop gap" until sufficient information and computational ability exist to build a more detailed and thereby more complex deterministic model. It is stochastic in the sense that in a finite time horizon only a limited number of contacts can be made out of the thousands of possible contacts. While there is a necessary and sufficient reason for each actual decision, these are to a degree indeterminate relative to the general principle that spatial diffusion reflects individuals' acquaintance fields. In other words, we do not need to know each person's precise contact network (the tree) to understand what typical diffusion processes or behaviors look like (the forest). Indeed, such details may obscure the identification of what may turn out to be a fairly simple yet elegant process.

The question of the degree of randomness of spatial diffusion is partly a matter of scale, a dilemma of whether the process should be viewed from micro- or macrovantages. If one is concerned with predicting individual or small-group behavior, an appropriate micro-level model can evaluate the probability of making a particular decision or contact out of many possible decisions. If interest centers on the aggregate behavior of a whole population, then a macro-level model can generate the general pattern of decisions and contacts. A simple diffusion example may help illustrate how the same model can be used both in a stochastic manner at the microscale and in a deterministic manner at the macroscale.

Consider the spread of a fad, such as high school students using the phrase "pig-out" to indicate an eating binge. At the micro or individual level, time and space constraints strictly limit the number of contacts possible, and any attempt to model such individuals requires a random choice allocator, such as a Monte Carlo procedure. Several formulas may be assumed, one that expresses the probability of adoption on the basis of individual characteristics, including their friendship networks and accustomed routes of movement. But we could also treat this as a macro or aggregate diffusion problem. Given the distribution of a sizable high school population, say 50,000 in a region, let a student in place x adopt or create the innovation. Spatial interaction formulas can be employed to allocate contacts deterministically rather than probabilistically. Suppose there are ten destinations, and the probability of contact between place x and these destinations are known, say P_i, i = 1, ..., 10. If the probabilities are viewed as proportions of N contacts, then the expected number of contacts at each destination is the product of NP_i. Hence if N = 100 contacts are to be made, and the probability of contact between place x and place i = 4 is 0.21, then the expected number of contacts at that place is 21.

Similarly, the contact patterns of these N = 100 adopters can be determined as a function of the distribution of opportunities around them. The ability to use a deterministic model is based on (a) large populations and (b) large numbers of contacts per person. Had we used the individual contact approach employing much shorter time periods *via* Monte Carlo decisions, the patterns of contact that result could converge at later stages of the diffusion process and thereby make stochastic and deterministic results appear as identical. This will happen because the probabilities do reflect the distribution of the population.

This finding means that (a) if we are interested in later stage macropatterns of diffusion of phenomena, then it is far simpler and more efficient to use a deterministic approach; (b) if we are interested in the patterns that result from decisions made by individuals, especially at early stages of the diffusion process, then a stochastic model is more appropriate.

Hierarchical Diffusion

In his earlier 1952 study, "Propagation of Innovation Waves," Hägerstrand determined that some phenomena such as radio tended to be adopted first in the larger cities, and then diffused down to smaller places. In the 1960s, as scholars experimented with spatial diffusion frameworks on a variety of phenomena, it soon became apparent that some diffusion phenomena could not be described as moving contagiously through a rural population. Rather, some phenomena involved a high threshold cost or market size for acceptance. Thus some researchers considered the possibility that there were patterns of diffusion other than contagious, such as hierarchical. The word *hierarchy* was originally used to rank groups of angels. It has been generalized to refer to the tendency for leadership of organizations or places in a settlement system to be arranged in a top-down linked set of graded ranks. In hierarchical diffusion the innovation may diffuse down the hierarchy of central places. This occurs for two main reasons: First, larger places may be more innovative and have more risk-taking persons; second, larger places have a greater potential for interaction. Because the spatial patterns between contagious and hierarchical patterns are so different, researchers initially considered the two diffusion patterns to require a totally new model framework.

In time, hierarchical diffusion has become recognized as a limiting case of a more general diffusion process that encompasses both the classifications of contagious and hierarchical diffusion. In examining the properties of pure hierarchical processes, Hudson (1969, 1972) recognized that it was not necessary for the actual structure of human contact to be either hierarchical or contagious. Rather, Hudson (see also Gale, 1972) found that diffusion fits within the general theory of spatial interaction. The theory of spatial interaction suggests that the likelihood of interaction is directly related to size of place and inversely related to distance between the point of interaction and the place. The resulting trade-off between place size and distance may result in a nearby medium-size city have a probability of contact equal to a much larger distant city. Pred (1971) demonstrated that diffusion could flow across similar levels of hierarchy (in keeping with broader notions of spatial interaction) versus the initial conception that constrained it to defined links in other ranks.

This duality between the pure forms of contagious-type diffusion and hierarchical-type diffusion was understood by Hägerstrand and provided him with the theoretical basis for his study of the diffusion of Rotary clubs in Europe (Hägerstrand, 1965). As part of the theoretical presentation, Hägerstrand presented an intuitive and now classic graphic depiction of this duality (Figure 4.6). In Figure 4.6, Hägerstrand suggests that the likelihood of contact, or expected amount of diffusion, might be considered at three spatial scales: (1) between a few very large national centers, (2) among sets of medium-sized regional centers, and (3) from local, national, or regional centers (contagiously) to much smaller nearby places. Hägerstrand believed that both processes, hierarchical and contagious diffusion, took place simultaneously.

Some types of phenomena require actual physical contact for their spread, such as some diseases: If the settlement density is even and travel local, then the process of contacting one's acquaintances around them will result in a locally contagious spread. However, certain financial tips may be more likely found among the managerial elite; if their distribution is highly clustered (metropolitan), and contacts often within branches of national firms, then the same basic process of contacting one's acquaintances will result in a hierarchical top-down spread.

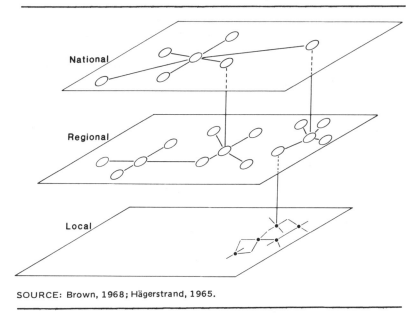

SOURCE: Brown, 1968; Hägerstrand, 1965.

Figure 4.6 Schematic Portrayal of Diffusion Viewed at Different Spatial Scales

Several decades of research have therefore resulted in the same conclusion as proposed earlier by Hägerstrand: Hierarchical or contagious processes do not represent different processes at all, but are different spatial manifestations of the same processes. Different patterns emerge because (1) phenomena have varying requirements for survival, including different critical masses or minimum market sizes, and (2) settlement systems differ in their concentration or degree of urbanization.

Spatial Diffusion and Spatial Interaction

The spatially diffused phenomenon must move away from an origin, whether contagiously or hierarchically, otherwise it is not diffusion. If contagious and hierarchical diffusion are different spatial manifestations of the same underlying process, then it follows that different models are neither needed nor appropriate. It is the classic spatial interaction models that best measure the probability contact fields and that, when used in a dynamic framework, best reproduce or predict patterns of spatial diffusion.

One of the more important contributions to our understanding that diffusion through a city system could best be viewed as an outcome of spatial interaction was in the work of Pederson (1970). He demonstrated that one could predict the timing at which an innovation would reach a place at a given level in the central place hierarchy, and at a particular relative spatial location (see Chapter 5). Berry (1972) verified that one could predict the timing of a given level of innovation adoption in his study of the diffusion of television in the United States; in this work Berry demonstrated the utility of the spatial interaction framework toward encompassing both hierarchical and contagious outcomes.

Spatial interaction models have also proven their usefulness in a variety of settings, including forecasting patterns of communication, work and shopping trip behavior, and the distribution of population and other activities. However, spatial diffusion as a process over space and time is more than merely one further example of spatial interaction models. Spatial interaction describes the normal *pattern* of contact among individuals, while spatial diffusion is the *process* of adoption or change that may result from those contacts when something "new" originates at a particular location. What happens at a farther distance is dependent on what happened earlier in time and at other closer locations.

A spatial interaction formula in a setting calibrated for diffusion can give the probability of contact from any location at any point in time, but this locus of greatest intensity of contact and change moves outwards from origins over time. Information requirements for the spatial interaction model calibrated for a diffusion problem include:

(1) the numbers and susceptibility of the population;
(2) the spatial distribution of the population; and
(3) the rate at which knowledge or opportunities to interact declines with distance.

At the same time, whether the phenomenon being modeled actually becomes diffused depends on the strength and attractiveness of the phenomenon and the "fervor" with which the early adopters spread the word. Experimentation (Morrill & Manninen, 1975) corroborates that the volume of contacts is absolutely critical to whether the process dies, spreads moderately or saturates a population.

The rate of distance decay of the contact field can be different among groups of individuals depending upon their social class, income, education, occupation, and even location. In studies of epidemiology (Kendall, 1965), there has been found "random mixing" within the home, and to a degree at the workplace or school, but a strong distance decay for other interactions. Random mixing may represent a state of affairs where individuals in the same group have significantly different contact fields; the contact field may even be different for an individual depending upon the type of phenomenon being diffused.

Several studies have proposed variations on the general spatial interaction theme as used in diffusion studies (Chapter 2). Gale (1972) relates the Hägerstrand concepts to Bailey's (1968) epidemiological model, and argues that Hägerstrand's model is a special kind of Markov process, and that the MIF contributes to Hägerstrand's model being more than descriptive. A Markovian process means that an outcome at one period is uniquely dependent on the situation of the immediately prior period. However, this conception of the diffusion process has not proven successful.

Yapa (1975) suggests that the "average simulation" map of a Hägerstrand model need not be derived by many replications, but can be determined from the MIF itself in a few steps. This is a variant of the themes of Berry (1972), Pederson (1970), and Webber (1972), and may be valid for reasonably deterministic macro-level situations.

While an "average map" may indeed be generated, this may not be meaningful or helpful in understanding actual micropatterns of adoption. In the absence of more detailed knowledge of determinants, an "average map" reduces to an uninteresting reproduction of the potential surface itself. The potential surface is a map of the relative accessibility of any point to the entire population around it, based on size and distance. This is an important point. The contact field is simply a reflection of the distribution of population; little is learned about diffusion as a process by fixing on a final pattern of acceptance that simply mirrors the pattern of contacts. What is of greater interest is the spatial variations in the unfolding of patterns on the way to a final more predictable one. It should be possible to construct a "variance map" that indicates the amount of deviance from the norm and that could be more informative and useful than the "average map."

Webber (1972) and Webber and Joseph (1978, 1979) develop an experiment with a variety of mathematical formulations and tests of diffusion processes, including derivations from physics and epidemiology. Webber constructs an electronic analog where nodes are represented as capacitors and corridors represented as resistors. The resulting electronic analog is tuned using hypothetical data from New South Wales, Australia. Webber's analysis of Hägerstrand's original data using his electrical analog demonstrates a spatial and temporal process that occurred at a rate different than in Hägerstrand's simulation.

Differential equations of a diffusion wave from epidemiology are derived, for which the Hägerstrand model is shown to be a solution. The importance of relative accessibility as in Pederson (1970) and Berry (1972) is explored by Webber, and the expected behavior of parameters is tested. It is shown that if one knows the rate of distance decay and the rank-size of places, then one can predict the time of arrival of the innovation (see Gale, 1972; Yapa 1975).

Morrill and Manninen (1975) determined that one simple diffusion model could replicate with appropriate empirical values for the parameters all the diverse patterns existent in the literature. The model was sparse in its requirements, needing only

(1) a spatial interaction contact field;
(2) a resistance to adoption being a function of the proportion of the population already having adopted, as in a logistic S-type function;
(3) variable levels and persistence in making contacts.

Morrill and Manninen's model is robust in that it can generate patterns that are hierarchical, contagious, or mixed hierarchical/contagious; low or high levels in population saturation; or "patchiness" (see Pielou, 1969) reflecting concentrations of susceptible populations. The pattern that is created has been demonstrated to be especially sensitive to the population distribution and the degree of distance decay in contacts. The model does not, however, explore the cases of partial or complete propagator control (Brown, 1981).

The Role of the Propagator:
The Market and Infrastructure Perspective

Emphasis in geographic work during the 1960s was upon the *pattern* of spatial diffusion and the *pattern* of final adoption by consumers; in other words, the emphasis was upon the *demand side*. Acceptance of an innovation by a firm, farm, or individual is generally directly correlated with wealth, highly educated decision makers, and decision makers that are risk takers. The richest, however, may be low risk takers therefore negating their propensity for innovation adoption.

In the late 1960s attention began to be redirected to sources of information such as electronic and newspaper media. The literature indicated quite clearly that spatial diffusion, in addition to being *passive*, could also be *active* and purposeful, namely, *propagated*. Emphasis in the research shifted to the *supply side*. Knowledge of diffusion processes could be used to obtain some economic or social goal, such as generating demand for a commercial product or obtaining population acceptance of family planning. However, not much was written in geography until the late 1960s in this area. Knowledge of the strategies of diffusion propagators may contribute to explaining why differences between actual and theorized patterns of diffusion may occur.

One important contribution to the literature on propagation of diffusion phenomena was by Brown (1981). Brown stressed the role of infrastructure and the importance to innovation of pursuing profit using a purposeful strategy of diffusion. Successful diffusion involved elements more complex than mere passive acceptance: A great number of diffused phenomena required propagators or entrepreneurs as proponents whose aim was to maximize the pace and spread of its acceptance. Brown concluded that (1) the origin of innovations was critical to their successful diffusion; (2) deliberate early establishment of centers of diffusion can increase an innovation's chance of being successfully diffused—in contrast, Hägerstrand took such centers as given; (3) deliberate strategy for spread might channel the phenomena into a pattern not otherwise taken.

The complex origin of phenomena, especially of innovations, is a multifaceted question in location theory; studies have most notably been done by economic historians such as Freeman (1982), Hall and Markusen (1985), Kelly et al. (1981), Mench (1979), Oakey (1984), and Sahel (1981) on this issue. One theory is that innovations are likely to occur in centers of relatively great size and power, places at the top of the central place hierarchy. Pederson (1970) suggests that the bias toward high-order places is because of the concentration of capital and research and development activity. However, actual experience does not support Pederson's hypothesis; rather, as demonstrated by Malecki (1977), in the United States, the highest rate of innovation occurs in intermediate-sized regional centers, and in cities faced with traumatic change. New England is cited as particularly successful in new activities replacing the declining textile industry.

The propagator may have a deliberate strategy for diffusion not necessarily tied to the short-term maximum profit motive of the entrepreneur, or maximum-satisfaction motive of the consumer. This strategy may or may not follow what would be predicted from a pure spatial interaction diffusion model of adoption. For example the rate of expansion may be regulated to maintain control or because of capital constraints.

In general, location theory formulations may be reasonably predictive for general industry trends, though they are usually not intended to be predictors of locations for specific firms. Location theory has been highly useful in prescriptive applications, that is, using various techniques to determine a set of optimal locations for a firm to choose between in a decision process. See Webber (1984) in this series for a review of industrial location. These theories are tied to principles of profit maximization or satisfaction maximization. The diffusion of outlets among firms that use applied geographic location theory will then follow the criteria applied by the geographer, rather than some other process at work in society. In this manner the geographer can act as the propagator responsible for location-diffusion patterns.

In highly urbanized societies or regions, the optimum location to begin a promoted diffusion is often at the top of the hierarchy. An important exception reflects the "neighborhood" effect where centers remain close to the point of origin to maintain effective managerial and quality control. A variation on this theme is for centers to expand to places similar in character to the origin; this is particularly important to the growth of franchises (Brown, 1981). The cautious, though guided, spread of Friendly Ice Cream is an example of this. Even over a period of 38 years, and despite name recognition and good profits, expansion was deliberately slow and spatially restricted to the northeastern portion of the United States.

Several generalizations have been observed in locating sites depending upon the management structure of the firm. If corporate decision making is centralized, a key strategy for determining location sites is at places where market potential or profit will be maximized. Such decisions can be modified by concern for access and control, thereby leading to combinations of hierarchical and neighborhood effects. If decision making is decentralized, a more dispersed pattern results with a tendency toward places with the highest market potential. But this tendency may be offset by the greater importance of interpersonal information links. These links tend to cause location patterns of a neighborhood character. Webber (1972) has presented evidence supporting a hypothesis that where decision making is decentralized, media influence is high, and personal influence is low, then adoption will be fast and spatially uniform; at the same time, firms in that type of environment will also be characterized by greater propensities for failure and retrenchment.

When firms require specialized infrastructures, then diffusion of such firms will be channeled to those locations that have invested in the infrastructure. Such specialized infrastructure investments usually require that local government officials and other community leaders be sensitive to needs of entrepreneurs. For instance, Brown (1981) determined that spread of Prolas cattle feeding was channeled to places that had investment infrastructure to cover the high start-up costs of special feed supplements.

Brown's work on infrastructure is of particular importance to the Third World. The economies of less developed countries can be partitioned into the market sector, where persons are able to adopt technical innovations, and the subsistence sector, where the value of innovation may be recognized but resources are insufficient to utilize the innovation. Diffusion is then not restricted to passive processes; the dual roles of planning and propagators can be critical to diffusion.

Diffusion and Development: Diffusion as a Planning Tool

Diffusion concerns the spread of change from limited origins. Diffusion transforms people and landscapes; in turn, this process often makes subsequent impulses of diffusion more acceptable. It is a small leap to consider then the effect that diffusion has upon economic growth. Impulses for economic change diffuse outward from some center. If society does not consider the propulsive force sufficiently strong, then knowledge of diffusion processes can assist in increasing the impetus of change.

From the standpoint of the individual, change through diffusion can be destructive as well as constructive. Certainly the invasion of North America by the Europeans was to the Native North American the advent of destruction of their culture and their people. On the other hand, the Europeans did bring with them technology and institutions that allowed for an unprecedented economic growth. In terms of the general evolution of humanity and our social system, the general trend of change has been positive. Clearly, however, centuries may pass where change is either negligible or destructive. For the individual, there is no certainty that diffusion processes will result in higher levels of welfare.

Diffusion of development impulses may be dispersed from a few centers; the growth of those centers may be at the expense of growth in other regions or in other parts of the center's region. For example, the growth of the late nineteenth-century northeastern region of the United States is in part attributable to regional exploitation of the post-Civil War South, and the new territories in the West. The shift in industry from the Snowbelt in the United States to the Sunbelt is a contemporary example of the economy of one region draining the economy of another. At more local scales, a similar process has occurred in the city. First a growing city may drain the nearby countryside of labor and capital; later spillovers from commuting and investment may revitalize the surrounding countryside. The role of Atlanta is a good example. Before 1960 there was depopulation in the outlying counties, but in recent years this trend has reversed (Morrill, 1974).

Many geographers have criticized the reliance upon diffusion impulses for economic development (Blaikie, 1978; Gaile, 1977, 1979). Such reliance leads to dependence of less developed countries upon more developed countries without there being an equal reliance of developed countries upon the less developed countries. Many geographers believe that reliance upon diffusion impulses should be encouraged only in those cases where there is mutual dependence. This is also true for the local level, such as the imbalance in an economically depressed region's dependence upon a single firm for local employment (Brown, 1981). Such imbalances lead to increases in inequality, greater class distinctions, and entrenchment of the dual economies of market and subsistence.

Much of the criticism is that while diffusion may be useful in stimulating development, it ignores other processes that are not characteristically diffusionist and ignores the structural problems faced in development. Dual economies, for example, are often intertwined in one system. The temptation to refer to the "diffusion of development" as the answer to problems of underdevelopment overly simplifies the complex process of development.

Diffusion can, however, play a major role in new market-based development policies. Diffusion studies can not only identify barriers that essentially are market imperfections, but can also identify methods to increase knowledge and free entry to provide a better market structure.

Conclusion

This chapter has outlined some of the directions in which diffusion research has developed following the monumental work of Hägerstrand. The literature has developed along six main themes: (1) modifications to Hägerstrand's mean information field; (2) the development of alternative and broader diffusion models; (3) relating hierarchical and contagious forms of diffusion; (4) relating spatial diffusion and spatial interaction;

(5) recognizing the role of the propagator in innovation diffusion; and (6) assessing the use of diffusion in development planning. This research has served to extend the utility of spatial diffusion. In the following chapter, some of the models discussed in the previous chapters will be mathematically formalized and operationalized.

5. MATHEMATICAL APPROACHES TO DIFFUSION MODELING

Introduction

This chapter introduces the mathematical foundations of elementary diffusion models. To apply spatial diffusion models to actual problems, it is necessary to have a fundamental knowledge of their mathematical structure; such knowledge is especially critical to those whose goal is to extend our understanding of diffusion processes. This is especially the case for those who seek to identify which process of diffusion is at work, for the performance of models can be compared on the basis of their mathematical structure.

The step-by-step mathematical operational procedure of Hägerstrand's model has been shown in the previous two chapters. In this chapter, variants of the Hägerstrand model will first be introduced; these are classified into those formulations that are stochastic and those that are deterministic. Second, the basic ideas central to mathematical epidemiology will be reviewed. Third, general space-time models are presented; space-time models are generally constructed by either adding time to spatial interaction models or adding space to temporal models. These models have been demonstrated to be very effective in simulating typical diffusion processes. One of the significant benefits of these models is their ability to predict at what time and location a given level of acceptance will be achieved for the particular phenomenon being diffused. Fourth, space-time trend surface models applied to measures of patterns of diffusion will be summarized.

Stochastic and Deterministic Variants of the Hägerstrand Model

To a large degree we have avoided formal mathematical and statistical specification of the diffusion problem; this is why the Hägerstrand model was presented earlier as a pedestrian step-by-step approach. Instead, here it is now presented in a more mathematically formal setting so that the stochastic and deterministic versions can be contrasted.

The stochastic variant of Hägerstrand's model can be stated as follows:

Let there be say twenty-five grid cells, each cell identifying the location of an adopter or set of adopters. The probability of a contact from origin i, defined as harboring an existing adopter, to any destination j, defined as harboring a susceptible or potential adopter, may be determined by the spatial interaction formula:

$$k \sum_{j=1}^{25} (A_i S_j)/d_{ij}^x = 1.0 = \sum_{\substack{j=1 \\ i<>j}}^{m} P_{ij}. \qquad [5.1]$$

The term k is a constant of proportionality. For each initial adopter i, or set of adopters at i, the probability p_{ij} of a contact with all possible locations is a direct function of opportunities, measured as the product of the initial adopter population A_i and susceptible population S_j, and inversely related to the distance d_{ij} separating locations i and j modified by x, an exponent of the distance decay function. Distance can be measured in many ways; often distance measured "as the crow flies" is inferior to other distance measures such as time or cost of the journey between places (see Haynes & Fotheringham, 1984).

Equation 5.1 defines a unique probability field around every location. This is a major departure from the classic *mean* information field. The allocations of probabilities are made in a manner similar to the Hägerstrand model that was discussed earlier. The probability field change for each time period reflects the change in the number of susceptibles S that remain. Probabilities for contact will increase as the numbers of adopters at a location rises; at the same time, contacts become increasingly "wasted" as the proportion of susceptibles to adopters declines.

The number and timing of contacts from an adopter may be separately specified using various statistical distributions, including the following example using a Poisson distribution:

$$n_t = k\, e^{-bt}. \qquad [5.2]$$

Both k and b are constants, their values are estimated statistically; e is the irrational nonrepeating number 2.71828. The number of contacts in period n_t is then an exponentially falling set of numbers, such as 8, 4, 2, 1. In the language of epidemiologists, this is known as a "removal" function or "death" relation for when an individual ceases to be a contactor. Barker (1977) has addressed the interesting question of the analysis of the "paracme" of diffusion, or the completion of the process of diffusion.

Some kind of diminishing rate of contact is a critical factor for most diffusions in controlling the extent of diffusion, and determining whether the phenomenon "catches on" at all.

Resistance on the part of potential adopters may be a function of the elapsed time of the process of whole diffusion, or, within a neighborhood, resistance may be a function of the proportion who have already accepted. These ideas may be formally expressed as:

$$R = \text{in} + [k/(P_{At}(U - P_{At}))] \qquad [5.3]$$

where int means the integer number; int[r] is then the integer of the required contacts closest to the value yielded by the expression to the right of the equal sign. Again, k is a constant, while P_{At} is the proportion of adopters at time t, and U is the ultimate maximum level of adoption.

Depending upon the value for the proportion of adoptors P_{At}, the proportion of adoptors at time t, equation 5.3 would lead to a sequence of required contacts like 3, 2, 1, 2, 3. Generally, in this type of resistance function, bandwagon effects will, through time, tend to increase the rate of adoption. For a classic discussion of bandwagon effects, see Veblen (1899).

The degree to which susceptibles are resistant to the innovation may also be determined by whether the personalities of individuals conform to pioneering-like spirits or innovators, majority adopters, or laggards. This does not necessarily lead to a systematic effect on the ultimate spatial patterns, but personality types do tend to expand the "crest" or zone of change, and will have a very noticeable influence on the generation to generation pattern (refer to Chapter 4 for discussion for resistance).

It is possible to construct a deterministic model of the spatial diffusion of populations if the locations being studied have large populations. The interaction formulation of equation 5.1 can be interpreted as describing the proportion of a large number of contacts from place i to all other locations j, rather than the probability of an individual contact to each location j. The model can be viewed as deterministic since the larger the number of contacts, the more closely the pattern reflects the actual distribution of opportunities around every location.

Another advantage of large numbers of contacts at most locations is that the population can be segmented to deal with different subpopulations. It may be that the diffusion pattern is different for those persons that are more spatially mobile *versus* those persons that are more spatially confined.

Iterations can be performed in a deterministic simulation as well as in a stochastic simulation, with probabilities and frequencies changing for each time period. The difference in focus between the two classes of simulation models is that in the deterministic model the behavior of the group is simulated, whereas in the stochastic model, the behavior of the individual is simulated. Hence unlike the stochastic model, there is no need for individual contact allocation in the deterministic model.

The easiest-to-understand spatial-temporal model of what unfolds over space and time is merely a sequential reapplication of a standard spatial interaction model where who is an adopter and who is a nonadopter is updated between time periods. Consider the following brief example (Table 5.1), which assumes an innovation begins in Pullman, Washington, and then diffuses across the state of Washington.

In time period O, Pullman has 100 adopters; the frequency of contacts as a function of distance and population was assumed to be given by the formula

$$F = 500(A_i S_j)^{0.5}/D_{ij}^2. \qquad [5.4]$$

In the first time period, 100 persons in Pullman interact with all other places, but to keep the model simple, the contact is recorded as effective only when the value of F exceeds 100. Only two links are effective: 800 more are contacted in Pullman itself, and 430 are contacted in the nearest large city, which is Spokane, Washington. In the second time period, the 800 new adopters in Pullman and the 430 adoptors in Spokane are sufficient to spread the innovation to five additional places, as well as sufficient to expand the adoption within both Pullman and Spokane. Once Seattle is reached in the second time period, its size leads to its dominance of the diffusion process. Here then there is a combination of hierarchical diffusion (Pullman to Seattle and Tacoma and thence downward in the city hierarchy), with that of local contagious diffusion (Pullman to nearby Walla Walla, Walla Walla to nearby Tri-Cities; Seattle to neighboring Tacoma). At the same time, even after four time periods have elapsed, the distance decay of contact

TABLE 5.1
Accumulated Adoption Over Time: Example Problem

Place	Population	Time 0	Time 1		Time 2		Time 3		Time 4	
					Number and Percentage Adopted					
Pullman	25,000	100 .004	900	.036	3,450	.14	8,650	.35	17,330	.69
Walla Walla	35,000		430	.0015	200	.006	3,050	.09	12,610	.36
Spokane	270,000				7,130	.026	32,510	.12	79,800	.30
Tri-City	110,000				470	.004	6,280	.06	26,530	.24
Yakima	80,000				110	.001	2,780	.034	14,350	.18
Seattle	1,400,000				370	.000	19,390	.014	164,390	.12
Tacoma	400,000				100	.0002	9,950	.02	87,250	.21
Vancouver	130,000						350	.003	5,740	.04

Distances	Pullman	Walla Walla	Spokane	Tri-City	Yakima	Seattle	Tacoma
Pullman							
Walla Walla	115						
Spokane	77	158					
Tri-Cities	133	50	135				
Yakima	193	132	192	85			
Seattle	285	263	275	215	145		
Tacoma	300	287	290	240	155	33	
Vancouver	343	257	343	210	180	163	134

NOTE: $F = 500\sqrt{\dfrac{A_i S_j}{D_{ij}^2}}$, where the minimum "internal distance" is 31.5.

For example, in Time 1: $F_{\text{Pullman-Spokane}} = 500\sqrt{100 \times 270{,}000 / 77 \times 77} = 438.198$;

Time 2: $F_{\text{Pullman-Seattle}} = 500\sqrt{800 \times 1{,}400{,}000 / 285 \times 285} = 206.011$.

Note that for simplicity, only flows over a 100 threshold were recorded, and then become new centers for diffusion.

remains a clear feature of the diffusion process; the evidence for this is to be found in the uneven proportions of adopters, namely 69% of Pullman, where the process started, versus only 12% in Seattle, and 4% in Vancouver, the town farthest from the others that have adopters.

Both the deterministic and stochastic variants of the model are useful in illustrating the process of spatial diffusion from one time period to the next. As the process of diffusion continues over successive iterations, each iteration symbolizing a time period, the results will merge with that of the actual population distribution. The stochastic version alone reveals the variety of microscale unfoldings that can occur as the diffusion process unfolds. Neither model can be considered as entirely satisfactory since time is not an explicit variable in either the stochastic or deterministic model. Time in equations 5.1 through 5.4 is essentially a by-product of the repetitive iterations of static spatial interaction processes. It is not then possible to predict with either model such desirable features as when certain levels of adoption will be reached in specific places. To do so requires a general interaction diffusion model.

General Interaction Diffusion Models and Critical Measures

Early theories of spatial diffusion were based on the gravity and spatial interaction model, which itself first came into geography as a physics analogue of a social process (Haynes & Fotheringham, 1984). Therefore, early pioneers in the field naturally looked again to the physical science literature for existing deterministic models of diffusion where its variables could be redefined; the revised model could subsequently be brought into the social science diffusion literature, as had their gravity and spatial interaction forebearers. Beckman (1970) borrowed from physics, and Haggett, Cliff, and Frey (1977) borrowed from epidemiology.

A second approach was to examine whether the spatial interaction concept could be made expressly temporal and thereby derive directly the information essential to a geographer such as the level, location, and timing of adoption. This was the motivation behind the work of Hudson (1969), Pederson (1970), Berry (1972), Casetti and Semple (1969), Gale (1972), Yapa (1975), Webber (1972), Haining (1983), and others. Their goals were to develop expressly dynamic models of human geographic diffusion processes. Before examining these ideas in more detail, a brief review of relevant formulations from epidemiology will be useful.

Epidemiology Models

In epidemiology, the population is divided into three groups: (1) *infectives*, who communicate a disease; (2) *susceptibles*, who are at risk of being infected; and (3) *removals*, who by way of death, recovery, immunization, or isolation can no longer give or receive the disease. Scholars who have worked on epidemiological diffusion models have been most interested in the time rates of infection, time rate of removal, and how these rates can initiate or terminate an epidemic. Mahajan and Peterson (1985) provide a comprehensive review of such temporal but largely aspatial models; spatial analysis enters their work by way of their interest in how population density and population mobility influence the likelihood, severity, and duration of an epidemic.

It was observed by several geographers that the logistic "S" curve, which is central to most epidemiological models, implicitly requires the fundamental assumption of free mixing: Contact is made by way of a "random walk." The free mixing of the epidemiological model would seem at first glance to be inconsistent with the geographers' idea that information behaves in a nonrandom spatially determined manner of distance decay from its source (Cliff & Ord, 1975; Hudson, 1972); moreover, the central geographic idea of neighborhood effects would require contacts to be spatially clustered and therefore nonrandom events. It was then argued that because of spatial considerations, the random walk model of diffusion was internally inconsistent with the nonrandom geography of information flow. Those concerned with reconciling the models responded to this challenge by showing that random local contact within a nonrandom pattern of information still met the condition for the logistic curve (Morrill, 1970). Out of these debates came the recognition that the geography of the landscape could and did affect the diffusion process. Kendall (1965) added quasi-spatial elements to the epidemiological models by including the effects of population density and the local proportion of removal to infected population.

Kendall assumed a three-part sequence to diffusion in his revised epidemiological model:

(1) The rate of conversion of susceptible to infected people (BP_{At}) is

$$BP_{At} = BDP_{st}A_t, \qquad [5.5]$$

The conversion rate depends on the rate of infection B, population density D, the proportion of total population that is susceptible P_{st}, and the average density of the already infected population in the immediate neighborhood A_t.

(2) The rate of conversion from infection to removals, gP_{At}, depends only on the proportion of population that is infected; this is an assumption that only the "natural course of the disease" is important.

(3) The net rate of change in the proportion of population infected R is equal to the rate of conversion of susceptible to infected less the rate at which the infected are removed from the population; namely,

$$R = (BDP_{st}A_t) - (gP_{At}). \qquad [5.6]$$

In the work of Kendall and others, an epidemic occurs if and only if D $>$ g/B or if BD $>$ g. This means that if population density exceeds the ratio of removals to infected in an area, then an epidemic wave will occur. The severity of the epidemic will depend on the degree of imbalance between the rate of infection and rate of removal. The higher the population density and local density of infectives, the more severe and widespread will be the epidemic and the greater will be the rate of change from susceptible to infective.

Bailey (1957) estimated that if DB / g = 1.4 then S = 0.5; namely, 50% will get the disease. If DB / g = 2.0 then S = 0.8. And if DB / g = 4.0 then everyone will get the disease. However, the reader should note that much of the result depends upon the *measure of density*. Density is one of the components that geographers have developed expertise in explanation and measurement. Because of the importance of a precise and accurate understanding of density, the expressions that geographers have developed have become necessarily more explicit and complex (see Thrall, 1987).

General Space-Time Models

The problem of proper mathematical specification of diffusion is essentially one of adequately treating both time and space simultaneously. The Hägerstrand model deals inadequately with time, the epidemiological model deals inadequately with space. The task of treating both has been approached in two main ways: (1) by adding time to spatial interaction models, or (2) by adding space (or distance) to a temporal model (such as a logistic model).

A temporal-spatial interaction model has been suggested by Pederson (1970), Berry (1972), and Webber (1972):

$$I_{it} = \sum k \frac{(S_i A_j)}{e^{bd_{ij}}}, \text{ when } T_j > F. \qquad [5.7]$$

Here I_{it} is the amount of contact (information) in place i at time t; S_i and A_j are, respectively, the numbers of susceptibles and adopters; d_{ij} is the distance between places i and j. The ratio $S_i A_j / e^{bd_{ij}}$ is part of the family of spatial interaction equations; it is the product of the numbers of opportunities that the adopters have of interacting with susceptibles, diminished by a weight that is an exponential function of distance. T_j is the time since the first contact was made, and F is the known threshold amount of time that must elapse from initial contact to first adoption.

The usefulness of equation 5.7 is in its ability to predict the time that a specific place will, if ever, receive the contacts necessary for subsequent adoption; moreover, places can be classified relative to others within a hierarchy of cities (King, 1984, p. 64). Essentially, the higher the potential interaction of a susceptible population, S_i, with other populations A already known to have acquired the innovation, then the sooner will the information threshold necessary for adoption F be surpassed. When places are cities within an urban hierarchical system, the values of I can be used to predict the rank-ordering of when the innovation will be adopted at a specific place; in this context the model can be interpreted as having spatial and temporal dimensions.

Spatial-Temporal Models

Several researchers approach a general spatial diffusion model by adding on a measure of space to a basic temporal model of diffusion. For example, refer to the work of Casetti and Semple (1969), Cliff and Ord (1975), and Haining (1983).

Casetti and Semple essentially adopted a variant on the logistic work. In their formulation, the proportion of population adopting at time t was $P_{At} = 1/(1 + \exp[a + bt])$. Space was included in their formulation essentially by arguing that the value that the estimated parameters a and b took on at a particular site depended upon a polynomial expression of distance between the site and the origin of the innovation: $a = a_o + a_1 d + a_2 d^2$, $b = b_0 + b_1 d + d_2 d^2$ (see also Casetti, 1972). The final model where parameters can be calibrated with appropariate data is then:

$$P_{At} = 1/(1 + \exp[a_0 + a_1 d + a_2 d^2 + b_0 t + b_1 dt + b_2 d^2 t]). \qquad [5.8]$$

Parameters of equation 5.8 are estimated using generalized least squares; the result is a "space-time trend surface." P_{At} now becomes, in equation 5.8, an estimate of the proportion of population adopting the innovation at distance d and time t from the place and time of origin. The nonlinearity of space is brought into the model by values of the parameters a_2 and b_2. The nonlinearity of time could have been accommodated in the model if the initial logistic expression were instead $P_{At} = 1/(1 + \exp[a + bt + ct^2])$ (see Casetti, 1972).

Casetti and Semple used data on the adoption of tractors in the midcontinental portion of the United States. In the statistical analysis, which parameters in equation 5.8 were statistically significant can be interpreted as a test for the validity of adding a spatial component to a logistic model. In their study, the significant variables were t (time), and d^2 (square of distance from an assumed origin in North Dakota). The dominant process was then logistic over time, but the level of adoption, or place on the logistic curve for a particular state, varied inversely with the square of its distance from the origin. In their study, by disaggregating spatial and temporal components, the temporal effects were found to be dominant for their particular data set; this, however, may be the by-product of using states as the base unit for spatial measurement. As a data unit, states may not show sufficient spatial resolution for this type of study.

Cliff and Ord (1975) use an auto-regressive model for forecasting. In their model, the rate of adoption depends on what happened in the previous time period. Specifically

$$r_t = bp_{t-1} + b_2 W p_{t-1}, \qquad [5.9]$$

where p_t is the proportion of population adopting at time t. The rate of change in adoption r_t varies directly with the cumulative proportion having adopted in the preceding period p_{t-1}, and on a potential measure W of the level of interaction of a state with its neighbors. In this manner, W will be high when levels of adoption in nearby states are high. Interaction in Cliff and Ord's model is estimated through an exponential distance decay formulation ($W_{ij} = e^{-bd_{ij}}$).

The model proposed by Cliff and Ord is a descriptive model rather than a behavioral model with interpretable parameters. At the same time, the model is more than a space-time fit to the underlying process because in diffusion theory what happens earlier in space and time is a critical predictor of what happens next.

Haining (1983) proposed a space-time version of an auto-logistic model, namely, a logistic model where distance to nearest neighbor was weighted. The equation

$$Px_{i,t} = \frac{e^{a+bY_{i,t}}}{1 + e^{a+bY_{i,t}}} \qquad [5.10]$$

is the probability of the x_i-th person adopting at time t. It is assumed by way of the auto-logistic temporal aspect of the expression that the probability of adoption at time t is itself a function of time.

That is, in the Haining model, the probability of adoption depends upon the state of affairs in the preceding time period, and the potential for the x_i-th person's interaction with persons that have adopted earlier. The interaction in time period t with propagators in time period t - 1 was expressed by Haining as:

$$y_{i,t} = x_{j,t-1}/f[d_{ij}]. \qquad [5.11]$$

where $y_{i,t}$ is the measure of the potential for the x_i-th person to adopt at time t, which in turn depends upon relative location to previous adopters. The term $y_{i,t}$ can be interpreted as a distance-biased interaction modifier. The term $f[d_{ij}]$ can take on several alternative forms: (1) it has been defined as a negative exponential measure of the friction of distance; (2) it has been estimated by the number of adopters among six neighbors nearest to the adopter; (3) and it has been estimated by the sum of all the distance-discounted adopters around x_{it}. Empirical analyses have shown equally good results using either of the first two measures of $y_{i,t}$.

Haining's formulation identifies a common logistic pattern of adoption among all individuals or places while, at the same time, the timing of each place depends on its location relative to places that have adopted earlier. In the case of the six nearest neighbors weighting for $f[d_{ij}]$, for example, the logistic pattern will not even begin until at least one neighbor has adopted. The model is then well suited to simulating microscale processes, and transforms the general Hägerstrand diffusion model into a true space-time diffusion process. Within Haining's model can also be found a geographic refinement of the Kendall (1965) epidemiological model.

The Haining formulation includes components that are both logistic and from the family of spatial interaction models; however, for its evaluation the complete calculation of the diffusion process through successive time periods must be computed. In contrast, a strength of the Casetti-Semple model is that a simple prediction of a place's level of adoption at any time can be made directly by substituting the time period and distance from the origin into the equation where its parameters have been earlier estimated; at the same time, a significant weakness of the Casetti-Semple model is their overly simplified treatment of space in the form of distance along a ray extending through the place of adoption and a possibly incorrectly identified origin.

What is needed appears to be, as Haining recognized, an identification of diffusion processes being logistic over time, and at the same time include a spatial interaction component that can specify just when the logistic process begins for any place or area (as in the work of Pederson, Berry, and Webber).

One rather direct solution is to use a standard logistic:

$$P_{At} = u/(1 + k \cdot \exp(a - bt)). \qquad [5.12]$$

except that now t is measured in relative time as

$$t_i = \frac{MAX \sum k(A_j S_i / e^{bd_{ij}})}{\sum k(A_j S_i / e^{bd_{ij}})}, \qquad [5.13]$$

which specifies the relative time that the process begins at each place.

Equation 5.13 is a modification of equation 5.7. The expression is interpreted in the following manner. Each place or area has an initial potential for interaction with the surrounding world. The number of susceptibles at place i is S_i, A_j are the number of original adopters in place j, and $e^{-bd_{ij}}$ is the discounting of opportunities further away with the weight of a negative exponential function. In equation 5.13 the aggregate potential within all places appears in the denominator, while the potential of that place with the highest potential appears in the numerator.

The time lag t_i, for one place relative to the place where the process of diffusion began, is directly related to the ratio of the maximum potential of any place to the potential of place i. The origin place of highest potential begins at relative time 1; the higher the ratio, the longer will be the lag and the longer will it take either for the diffusion process to begin or for a given level of adoption to be reached.

One of the best and most direct measures of the way in which the spatial interaction component specifies the timing of the diffusion process is the Pederson-Berry-Webber formulation of equation 5.7; there the diffusion potential was calculated as a function of the interaction between susceptibles in place i and adopters in all other places j. A drawback of the Pederson-Berry-Webber formulation is that it may overemphasize the influence of the possible accidental location where a process started, as opposed to measuring the true potential of diffusion inherent as a characteristic of the actual population distribution. One solution to this problem of the Pederson-Berry-Webber formulation is to specify the time lag as functions of both the initial diffusion potential and the inherent population potential

$$t_i = \frac{MAX \sum k(A_j S_i \ P_i P_j / e^{bd_{ij}})}{\sum k(A_j S_i \ P_i P_j / e^{bd_{ij}})},$$

Equation 5.14 differs from the Pederson-Berry-Webber formulation in equation 5.13 by the expansion of the measure of potential interaction to include two components: the initial potential reflecting the distribution of early adopters ($A_j S_i$), and the inherent potential for interaction reflecting the total population of the origin P_i to all other places P_j. The effect is to cause the diffusion process to begin somewhat earlier in places with very high inherent potential, even if they are initially remote from early adopters. In actual case studies, the relative influence of these two kinds of potential can be measured.

To sum up, the following three elements are generally present in all space-time diffusion models:

(1) The logistic relationship is the most commonly observed pattern of acceptance, especially when adjustments have been made for the different times when the process begins at the various locations. Because of this empirical regularity, and the reasoning behind the relationship, a logistic "S" curve is often the structural form used when a specific equation must be assumed.
(2) In the most elementary or parsimonious models, the estimated time or intensity of the adoption process at each location is expressed as a function of both initial adoption potential and inherent adoption potential. More complex models add other factors hypothesized to be instrumental in the diffusion process, including various socioeconomic relationships.
(3) There is a clear dependency of the pattern that characterizes the acceptance of the phenomena over time and the: (a) the quickness of the adoptive process; (b) the duration of the adoptive process; (c) and the

differential in the timing of the spread of the adoptive process (see, for example, Webber, 1972).

Once the process of diffusion has begun, places thereafter receiving the diffusion stimulus may more quickly adopt the phenomenon for by then there will be more information and less uncertainty about the innovation. Nevertheless, if the phenomenon being diffused reaches a resistant subpopulation that considers the phenomenon not to be profitable or to be of little use, then the process of diffusion will proceed there at a slower rate.

Diffusion processes become clearer after temporal and spatial aspects of the process have been separated and distinguished from one another. As an illustration, consider again the studies of the diffusion of tractors by Casetti and Semple (1969), Cliff and Ord (1975), and Morrill (1985). In the process of diffusion, some variation will likely occur in the logistic pattern of timing as indicated above in aspect 3; in these studies, however, the rate of adoption was not systematically faster in late adopting states or slower in early adopting states. Analysis of covariance is an appropriate statistic to use in such settings when one wants to derive an estimate of the logistic pattern as in aspect 1 above, and depicted in Figure 5.1. The problem statement for such an analysis of covariance is as follows: The dependent variable is the proportion of population of each state adopting tractors. The independent variable is time. Measurements from each place (state) are identified by separate binary dummy variables; in this manner one can test whether the logistic relation between rate of adoption and time is the same, allowing for the fact that the stimulus may reach each state at different times. The dummy variable for each state is constructed by setting the variable equal to 1.0 if the observation occurs in the state and zero otherwise. This then allows for the estimation of a mean time lag for each state, thereby satisfying aspect 2 above. If the results of the analysis of covariance identify a regular geographic pattern of the time lags about the origin of the phenomenon, then a spatial diffusion process can be said to have been verified.

Casetti and Semple (1969) essentially estimated the lag as a function of distance from a North Dakota origin; Cliff and Ord (1975) used a spatial auto-correlation function of the sequential spatial pattern of adoption; Morrill (1985) expressed the lag as a function of initial potential, inherent potential, and various social-economic characteristics of the states. Therefore, if a common logistic trend can be identified, and if corrections can be made for the differential timing for when the

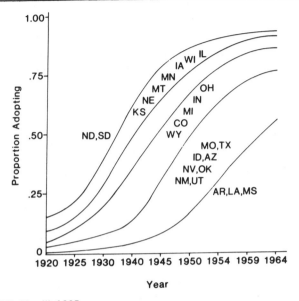

SOURCE: Morrill, 1985.

Figure 5.1 Cumulative Proportion Adopting Tractors: Means for Groups of States

diffusion phenomenon begins at the various sample sites, then the spatial component can be isolated and then analyzed.

Analysis of variance in the example of the diffusion of tractors in the midcontinental United States can be an appropriate test for demonstrating the existence of a common logistic pattern of adoption. Ordinary least squares can be successfully used to estimate the lag in years between when states reach a certain level of adoption; the lag can be estimated in terms of the initial potential (relation to nonadopters), inherent potential (reflecting the entire population distribution), and selected social-economic characteristics of the state. In Figure 5.1, groups of states with similar positions of adoption have been agglomerated to aid in the interpretation of the graph. The logistic growth pattern evident over time is revealed to be a dominant component of the adoption of tractors; at the same time there remains a significant lag, and identification of those states that have particular time lags by itself reveals an ancillary component of spatial diffusion.

The example shows that the "fate" of a place (time at which a given proportion of persons adopt) can be estimated by knowing initial conditions, the population potential, and by making realistic assumptions concerning the behavior of time parameters. This does not mean to say that it is a trivial matter (Haining, 1983) to estimate the parameters from known data, for there are distance decay parameters for estimating potential, there are parameters for the logistic expression, and most critical, additional parameters are required for the appropriate determination of relative time.

Measurement and Evaluation of Expected Diffusion Behavior

A variety of tests of diffusion patterns has been presented in Chapter 3. Here we are concerned with formal expressions of the diffusion of a phenomena in terms of distance and time, and to evaluate whether a characteristic of the process of diffusion is a wave-like spread across the landscape. A variety of different expressions can be used when the parameters of the model are not required to be interpretable in some behavioral context, but are used merely to assist in describing and perhaps predicting diffusions. The common trend surface is one such useful method that allows for the inclusion of absolute spatial location into the description of the diffusion process.

Absolute spatial location is usually measured by Cartesian (x, y) coordinates, or longitude and latitude. Relative spatial location measures the location of a place in terms of another place, for example, the relative location of a place can be measured in terms of road distance to another place.

A trend surface equation reproduces a topographic-like contour surface; the specific contour surface that is shown to unfold across the landscape is a result of the diffusion of the phenomenon across the landscape that in turn is reflected in the values of the parameters that have been estimated by the trend surface algorithm. Instead of the measure of altitude in a topographic contour surface, here the dependent variable F_{xyt}, is the number adopting or the percentage adopting at location (x, y) at time t.

Trend surfaces are represented by equations and one of the more useful is a polynomial where independent variables include the coordinates of the observations as well as time. A set of maps can then depict the contours showing the expected time when different proportions of acceptance have been attained at the various locations. The model is constructed by estimating parameters b for every combination of independent variables x, y, x^2, y^2, t, and t^2, such as in the following:

$$\begin{aligned}F_{x,y,t,} = &\ b_0 + b_1 t + b_2 t^2 + \\&+ b_3 x + b_4 xt + b_5 xt^2 + \\& b_6 y + b_7 yt + b_8 yt^2 + \\& b_9 x^2 + b_{10} x^2 t + b_{11} x^2 t^2 + \\& b_{12} y^2 + b_{13} y^2 t + b_{14} y^2 t^2 + \\& b_{15} xy + b_{16} xyt + b_{17} xyt^2 + \\& b_{18} x^2 y + b_{19} x^2 yt + b_{20} x^2 yt^2 + \\& b_{21} xy^2 + b_{22} xy^2 t + b_{23} xy^2 t^2 + \\& b_{24} x^2 y^2 + b_{25} x^2 y^2 t + b_{26} x^2 y^2 t^2.\end{aligned}$$ [5.15]

The greater the complexity of the surface, the more observations are required for statistical validity (Agterberg, 1984). For a discussion on the interpretation and derivation of such (x, y) surfaces see Thrall (1984) and Chorley and Haggett (1965). Casetti and Semple's (1969) "spatialized logistic" and Morrill's (1968) Poisson form suggested from wave theory are alternative approaches to the analysis of spatial-temporal diffusion.

Conclusion

This chapter has summarized the foundations of the mathematical expressions for spatial diffusion. Mathematical spatial diffusion theory was first seen in Hägerstrand's Monte Carlo simulation model, and has subsequently grown to include formulations that introduce time into a basic interaction equation, introduce relative spatial location by way of distance measurements, and introduce absolute spatial location by way of Cartesian coordinates. Forecasts of when a specific place is likely to attain a certain level of adoption can be made when time is included in a mathematical model of diffusion with measurements of either absolute or relative spatial location.

6. PRESENT STATUS, NEEDS, AND DEVELOPMENTS

Consider the following outline that can be used to classify contributions of spatial diffusion reasoning by the discipline of geography:

Classification by Broad Topic or Theme
    (a) Innovation adoption: social, economic, political, agricultural
    (b) Diffusion of innovation and cultural development
    (c) Epidemiology

(d) Settlement, migration, and landscape evolution

Classification by Methodological Approach
- (a) Descriptive
- (b) Parameter estimation, testing of theories and models
- (c) Mathematical model development
- (d) Stochastic/microscale
- (e) Deterministic/macroscale

In the first classification set, attention has been placed upon the category of innovation diffusion (see Clark, 1984), including the spread of new products, practices, and ideas. Within this tradition over the years, concern has shifted somewhat away from microprocesses, that is away from a study of the behavior of small numbers of persons, away from an emphasis on local information, and away from detailed studies of spatial movement itself. Instead, concern has shifted toward greater emphasis upon the behavior of large numbers of persons in aggregate, toward the role of such aspects as thresholds for adoption and infrastructure, toward the causes of variations in spatial movement, and toward the consequences of innovation diffusion on individuals and societies.

Research on innovation diffusion has come to be interdependent with that of economic and social development. Initially, this overlap came about because of case studies of innovation in the Third World setting. These Third World case studies quickly were recognized as contributing new ideas on issues and strategies of economic development. A cleavage soon thereafter appeared. On the one hand, there were those who viewed economic development as a successful diffusion process that integrated Third World economies with the new international corporate organization. On the other hand, there arose a largely anthropological or structuralist school with a viewpoint critical of an integrated world economy; it has been argued that such an economy would create an undesirable dependency of Third World countries upon the more technological culture of developed countries; the result would be a destabilization of the local economy of Third World countries, a loss of local control over the direction of economic development, and a destruction of Third World cultures and languages.

During the past three decades, there has been a continual interest in the diffusion of epidemics. To some extent this is purely pragmatic because of government funding and the general high quality of data. Epidemiological studies then have become ideal vehicles for verification of theory and continued development of mathematical models.

In part because of the relatively high level of resources available in epidemiological studies, and the relative lack of resources available for other areas of diffusion study, research has not developed to the extent that it otherwise could have. This is especially true of quantitative research. Breakthroughs that otherwise may have come from promising research in such areas as settlement diffusion, urban expansion, and growth have not then been forthcoming. This is especially regrettable in the context that this literature has a great potential for contributing practical strategies for economic growth of less developed places.

In the second classification set, methodological approaches, the tradition of descriptive diffusion studies of culture continues. This literature has the virtue of treating diffusion within a broad social context. A handicap of this literature is that these cultural case studies are not analytical and typically cannot be broadened into general theories. Hence their value is limited largely to the particular case study they describe and do not contribute insight into diffusion processes of different phenomena, or even the same phenomena but at different locations or times.

In contrast to the descriptive cultural studies, the enormous revolution that Hägerstrand is responsible for led to a rich body of work along two important paths. In the 1960s, there were many replications of his stochastic simulation model using a variety of data. Examples include studies on settlements and epidemics as well as the spread of ideas. This research led to attempts to measure such concepts as information and resistance. In the 1970s, scholars expanded the concept of diffusion to include hierarchical processes; deterministic mathematical models of diffusion were developed. This quantitative work led to a broadening of the general ideas that encompass manifold diffusion processes that are manifested in widely varying forms. It is now recognized that Hägerstrand's ideas are only a limited special case of diffusion. At the same time, Hägerstrand's seminal ideas remain in great historical prominence.

Concern for verification has led to interest in testing patterns expected from theory. Dissatisfaction among researchers in quantitative techniques has been expressed on the indeterminancy of Hägerstrand's model, and this has led in turn to the development of general mathematical statements the outcomes of which have greater predictability.

So what have we learned? What are the important determinants of spatial diffusion? What are the parameters that matter most? This will be answered in the context of Rogers's (1983) observation that five essential elements have come to be associated with diffusion: the phenomenon, communication, distance, time, and social structure.

Phenomenon. The nature of the phenomenon being diffused does matter, although this has received less attention than it otherwise deserves. Some questions that should be considered when investigating diffusion phenomena include:

(1) What are the characteristics of a phenomenon that make it attractive, and how attractive must it be to "catch-on"?
(2) Is it a short-run phenomenon such as a rumor or a long-run phenomenon such as settlement patterns? What is the behavior of the phenomenon over time and what is the role of the passage of time? How long will the effect of the phenomenon last?
(3) Does the phenomenon have broad or narrow appeal? How does this affect where the process will occur?
(4) What is the threshold of infrastructure requirements? How do differing thresholds affect who can adopt and where the process can occur (Brown, 1981)?
(5) What is the degree to which diffusion is controlled by a propagator (Brown, 1981) *versus* being brought about by a contact pattern among individuals? This is a factor that is far more important than originally thought.
(6) How independent is the phenomenon from other competing possibilities?
(7) What is the degree of dependence on personal contact, local knowledge, and/or commercial and public information media?
(8) Do more costly phenomena lead to greater or smaller lags in awareness of the phenomena, and greater or smaller lags between awareness and decision, than less costly phenomena? Is this lag different for different types of phenomena, and if so, what characteristics of the phenomena lead to differing degrees of lag?

The lower the population density of persons, and the further apart are clusters of more highly dense population, the slower will be the dispersion of the information, the lower will be the level of information flow, and the lesser will be the range over which information will flow. Communication among the principal actors of the diffusion process is the key to diffusion; important to the diffusion process are the propagators, and the media, information linkages between adopters and propagators, and the distance between the various actors. It is for this reason that the spatial interaction models (Haynes & Fotheringham, 1986) are so successful in forecasting diffusion since they make explicit the trade-off between size and distance. But diffusion requires more than just dispersal of information, for the information must be adopted, and there is no guarantee that those who become aware will also adopt.

Communication. This also plays a central role in the behavior of firms or propagators that deliberately set out to diffuse phenomena and to convince others to adopt (Brown, 1981). But just as information and individuals' travel declines with distance, so does the ability of a firm to maintain effective control decline with distance from its headquarters (Thrall, 1984). The decisions that firms make reflect this spatial pattern of control. The resulting spatial pattern remains one that maximizes individual utility or firm profits, but subject to the constraint of the spatial pattern of information and control.

Distance. The likelihood that individuals will accept the phenomenon after people become aware of the phenomenon is affected by the distance between these people. This understanding has come about not because geographers have biased the results to conform to their own *raison d'etre*; rather, oddly, geographers have regularly outlined their research curriculum with a personal goal toward proving that distance does not matter. Their results clearly show that indeed distance does matter.

Time. The role of time in diffusion is, for the large part, more complex than that of distance. Time, the period through which the diffusion phenomena will exist, influences the probability that new phenomena will emerge to compete with the earlier diffused phenomena, and influences the strength of the diffusion impulse. Time has a direct role in the eventual decline of evangelistic propagators contacting new adopters. Time is central to the eventual decline in resistance to the new phenomena; time then is inseparable with the rate of adoption and the rise in the proportion of adopters. These aspects of time affect the spatial trend and ultimately the spatial extent of the phenomena.

Social structure. It is now recognized that the complexities of social structure must be dealt with when calculating the likely success of the phenomena being adopted, and the very nature of its spread. First, the role of leadership and its impact upon setting threshold reuqirements must be understood. Second, it must be recognized that certain phenomena are targeted to selected subpopulations. Third, the phenomena being diffused may itself change the structure of society, as well as change the role that social structure has in subsequent planning and propagation of diffusion phenomena. All of these aspects can be studied, measured and evaluated, and modeled mathematically, though probably not as a consolidated and yet still simple framework.

There appears to be three analytic stages to modeling mathematically diffusion phenomena. The first stage is identification of the origin of the phenomena and the pattern of initial adoption; this is often viewed as a special case of the general literature on location theory. The second stage of the analysis measures the actual pattern of spatial diffusion, by way of temporal and spatial interaction/potential formulations. Variables concerning the phenomenon's attractiveness and population selectivity are included in the model. Parameters are measured that calculate the lifetime of the diffusion and the infrastructure or threshold requirements, the nature of the spatial interaction/potential field and the role of time with respect to contact frequency and resistance. The third stage evaluates the bias that is inherent in the model because of its mathematical structure alone, and the consequences that this has upon the type of landscape that is forecast. While the first and third stages are not "spatial diffusion" as such, how these stages are dealt with directly translates into how the final diffusion problem will be understood; it is then both a necessity and a responsibility for those working in mathematical diffusion theory to deal with these issues.

Needs and Future Trends

It is appropriate now, when bringing this book to a close, to devote some attention to the "pre" and "post" aspects of diffusion; that is, the geography of innovation and propagation behavior *versus* the consequence or aftermath of the diffusion processes. Viewing diffusion as "good" or "bad" illustrates limited sight, for what we are today is a collection of past diffusions; at the same time, because we are living in a world that is quickly changing, there is now a great need for scientific analyses of long-term impacts and trends, and to assess where, why, and how some diffusion processes succeed or fail, and for whom. As the diffusion impulse is carried through society, what will be the ultimate effect upon the well-being of that society?

Like that of any disciplinary subfield, one needs to appreciate the complexity of diffusion processes and to respect the diversity of approaches to the problem. It is fruitless to assert that either micro/behavioral or macro/deterministic modeling is a superior approach. Indeed, it may be that some diffusion processes can contribute to an understanding of how micro and macro approaches can be best integrated. Beginning with Hägerstrand, many scholars have blended the abstract theories of diffusion with empirical analysis. Still, however, we remain negligent in formal statistical tests with an aim to verifying the mathematical or theoretical arguments; it has seldom been the case in the diffusion literature that comparisons have been made between theoretical expectations and empirical observations.

There is also a need for explicit analyses of the stability of parameters within empirically calibrated models, and to forecast what events or conditions will lead to changes in the parameters. The variability and effect of parameter stability has long been a tradition in economics, but has not become of significant concern in geography, and particularly it has seldom been a component of analyses of spatial diffusion.

A broad picture of the concern of the general geographic literature has been on the effect that a particular phenomena has upon the landscape and the evolution of that landscape. Yet diffusion research in geography has clearly moved away from this central query of the discipline. Emphasis should return to the effect that diffusion phenomena have upon the landscape and the role of diffusion in building the landscape to be as it is. What is the role of diffusion in differing landscape patterns and their evolutions? To answer such questions, there must evolve a more quantitative historical geography and a more human and historical quantitative geography. This raises the issue of focused research objectives, and a commitment of resources greater than that which afford only narrow and often trivial diffusion case studies. Rather, the wider geographic picture should be painted of the meaningfulness of diffusion, a geographic composition where diffusion processes loom large.

REFERENCES

Agterberg, F. (1984). Trend Surface Analysis. In G. L. Gaile & C. J. Willmott (Eds.), *Spatial statistics and models.* Dordrecht: D. Reidel.
Bailey, N.T.J. (1957). *The mathematical theory of epidemics.* London: Charles Griffin.
Bailey, N.T.J. (1968) Stochastic birth, death and migration processes for spatially distributed populations. *Biometrika, 55,* 189-198.
Barker, D. (1977). The paracme of innovations: The neglected aftermath of diffusion or a wave goodbye to an idea. *Area, 9,* 259-264.
Bartlett, M. S. (1975). *Statistical analysis of spatial pattern.* London: Chapman Hall.
Barton, B., & Tobler, W. (1971). A spectral analysis of innovation diffusion. *Geographical Analysis, 3,* 182-186, 195-199.
Beckmann, M. (1970). Analysis of spatial diffusion processes. *Papers, Regional Science Association, 25,* 109-117.
Berry, B.J.L. (1972). Hierarchical diffusion: The basis of developmental filtering and spread in a system of growth centers. In N. Hansen (Ed.), *Growth centers in regional economic development.* New York: Free Press.
Blaikie, P. M. (1978). The theory of spatial diffusion of innovation: A spacious cul-de-sac. *Progress in Human Geography, 2,* 268-295.
Blumenfeld, H. (1954). The tidal wave of metropolitan expansion. *Journal of The American Institute of Planners, 20,* 3-14.
Boots, B., & Getis, A. (1988). *Point pattern analysis.* Newbury Park, CA: Sage.
Bowman, I. (1931). *The pioneer fringe.* New York: American Geographical Society.
Brown, L. A. (1968). *Diffusion dynamics.* Lund: Lund Studies in Geography #30.
Brown, L. A. (1981). *Innovation diffusion: A new perspective.* London: Methuen.
Casetti, E. (1972). Generating models by the expansion method: Applications to geographical research. *Geographical Analysis, 4,* 81-91.
Casetti, E., & Semple, R. K. (1969). Concerning the testing of spatial diffusion hypotheses. *Geographical Analysis, 1,* 254-259.
Chorley, R., & Haggett, P. (1965). Trend-surface mapping geographical research. *Transactions of The Institute of British Geographers, 37,* 47-67.
Clark, G. (1984). *Innovation diffusion: Contemporary geographical approaches* (CATMOG #40). Norwich: Geo Books.
Cliff, A. D. (1968). The neighborhood effect in the diffusion of innovation. *Transactions of The Institue of British Geographers, 44,* 75-84.
Cliff, A. D., & Haggett, P. (1979). Geographical aspects of epidemic diffusion in closed communities. In N. Wrigley (Ed.), *Statistical applications in the spatial sciences.* London: Pion.
Cliff, A. D., Haggett, P., & Graham, R. (1983). Reconstruction of diffusion process at local scales (measles epidemic in Iceland). *Journal of Historical Geography, 9,* 29-46, 347-368.

Cliff, A. D., & Ord, J. K. (1973). *Spatial autocorrelation*. London: Pion.
Cliff, A. D. & Ord, J. K. (1975). Space time modelling with an application to regional forecasting. *Transaction of The Institute of British Geographers, 64,* 119-128.
Cox, K. R., & Demko, G. J. (1968). Conflict behavior in a spatio temporal context. *Sociological Focus, 1,* 55-67.
Davies, S. (1979). *The diffusion of process innovations.* Cambridge: Cambridge University Press.
Dodd, S. (1950). The interactance hypothesis. *American Sociological Review, 15,* 245-256.
Edmonson, M. (1961). Neolithic diffusion rates. *Current Anthropology, 2,* 71-102.
Filipiak, J. (1983). Diffusion equation model of slightly loaded M/M/1 Queue. *Operation Research Letters, 2,* 134-139.
Forbes, D. K. (1984). *The geography of underdevelopment.* Baltimore: Johns Hopkins Press.
Freeman, C. (1982). *The economics of industrial innovation.* Cambridge: MIT Press.
Freeman, D. B. (1985). The importance of being first: Preemption by early adopters of farming innovations in Kenya. *Annals of the Association of American Geographers, 75,* 17-28.
Gaile, G. L. (1977). *A critique of the diffusion of development.* Paper presented to the annual meeting of the Association of American Geographers, Salt Lake City, Utah.
Gaile, G. L. (1979). Spatial models of spread-backwash processes. *Geographical Analysis, 11,* 273-288.
Gale, S. (1972). Some formal properties of Hägerstrand's model of spatial interactions. *Journal of Regional Science, 12,* 199-217.
Geary, R. C. (1954). The contiguity ratio and statistical mapping. *The Incorporated Statistician, 5,* 115-145.
Gilg, A. W. (1973). A study in agricultural disease diffusion: The case of the 1970-71 fowl pest epidemic. *Transactions of the Institute of British Geographers, 59,* 77-97.
Gould, P. R. (1970). Tanzania 1920-1963: The spatial impress of the modernization process. *World Politics, 22,* 149-170.
Griliches, Z. (1957). Hybrid corn: An exploration in the economics of technological change. *Econometrica, 25,* 501-522.
Hägerstrand, T. (1952). *On the propagation of innovation waves.* Lund: Lund Studies in Geography B No. 44.
Hägerstrand, T. (1967). *Innovation diffusion as a spatial process.* (A. Pred, Trans.) Chicago: University of Chicago Press. (Original work published 1953)
Hägerstrand, T. (1965). Aspects of the spatial structure of social communication and the diffusion of information. *Papers, Regional Science Association, 16,* 27-42.
Hägerstrand, T. (1967). On Monte Carlo simulation of diffusion. In W. Garrison & D. Marble (Eds.), *Quantitative geography.* Evanston: Northwestern University Studies in Geography No. 13.
Haggett, P. (1979). *Geography: A modern synthesis* (3rd ed.). New York: Harper & Row.
Haggett, P., Cliff, A., & Frey, A. (1977). *Locational analysis in human geography.* New York: John Wiley.
Haining, R. (1982). Interaction models and spatial diffusion processes. *Geographical Analysis, 14,* 95-108.
Haining, R. (1983). Spatial and spatial-temporal interaction models, and the analysis of patterns of diffusion. *Transactions of the Institute of British Geographers, 8,* 158-169.
Hall, P., & Markusen, A. (1985). *Silicon landscapes.* Boston: Allen & Unwin.

Harvey, D. W. (1966). Geographical processes and the analysis of point patterns. *Transactions of the Institute of British Geographers, 40,* 81-95.
Haynes, K. E., & Fotheringham, A. S. (1984). *Gravity and spatial interaction models.* Newbury Park, CA: Sage.
Hudson, J. C. (1969). Diffusion in a central place system. *Geographical Analysis, 1,* 45-58.
Hudson, J. C. (1972). *Geographical diffusion theory.* Evanston: Northwestern University Studies in Geography No. 19.
Jordan, T., & Rowntree, L. (1981). *The human mosaic.* San Francisco: Canfield.
Karatzas, I. (1984). Gittins indices in the dynamic allocation problem for diffusion processes. *Annals of Probability, 12,* 173-192.
Kelly, P. et al. (Eds.). (1981). *Technological innovation: A critical review of current knowledge.* San Francisco: San Francisco Press.
Kendall, D. G. (1965). Mathematical models of the spread of infection. In Medical Research Council (Ed.), *Mathematics and computer sciences in biology and medicine.*
Kimura, T., & Ohsone, T. (1984). A diffusion approximation for an $M/G/M$ queue with group arrivals. *Management Science, 30,* 381-388.
King, L. J. (1984). *Central place theory.* Newbury Park, CA: Sage.
Lamm, G. (1984). Extended Brownian dynamics: 3-dimensional diffusion. *Journal of Chemical Physics, 80,* 2845-2855.
Landes, D. S. (1969). *The unbound Prometheus: Technological change and industrial development in western Europe from 1750 to the present.* Cambridge: Cambridge University Press.
Lankford, P. (1974). Testing simulation models. *Geographical Analysis, 6,* 295-302.
Lansky, P. (1983). Inference for the diffusion models of neuronal activity. *Mathematical Biosciences, 67,* 247-260.
Lowe, J. C., & Moryadas, L. (1975). *The geography of movement.* New York: Houghton Mifflin.
Mahajan, V. & Peterson, R. A. (1985). *Models for innovation diffusion.* Newbury Park, CA: Sage.
Malecki, E. (1977). Firms and innovation diffusion: Examples from banking. *Environment and Planning A, 9,* 1291-1305.
Marble, D. (1967). *Some computer programs for geographic research.* Evanston: Northwestern University.
Marble, D., & Nystuen, J. (1963). An approach to the direct measurement of community mean information fields. *Papers, Regional Science Association, 11,* 88-109.
Mayfield, R. (1967). Spatial structure of a selected interpersonal contact: A regional comparison of marriage distances in India. In R. Mayfield & P. English (Eds.), *Man, space and environment* (pp. 385-401). Oxford: Oxford University Press.
McKeague, I. W. (1984). Estimation for diffusion processes under misspecified models. *Journal of Applied Probability, 21,* 511-520.
McMaster, D. N. (1962). Speculations on the coming of the banana to Uganda. *Journal of Tropical Geography, 16,* 57-69.
McVoy, E. C. (1940). Patterns of diffusion in the United States. *American Sociological Review, 5,* 219-227.
Meinig, D. (1965). The Mormon culture regions: Strategies and patterns in the geography of the American West, 1847-1964. *Annals of the Association of American Geographers, LV,* 191-220.
Mench, G. (1979). *Stalemate in technology.* Cambridge: Ballinger.
Metcalfe, S. (1981). Impulse and diffusion in the study of technical change. *Futures, 13,* 347-359.

Mollison, D. (1977). Spatial contact models for ecologic and epidemic spread. *Journal of the Royal Statistical Society, Series B, 39,* 283-326.

Moran, P.A.P. (1948). The interpretation of statistical maps. *Journal of the Royal Statistical Society, 10,* 243-251.

Morrill, R. (1965). The Negro ghetto: Problems and alternatives. *Geographical Review, 55,* 339-361.

Morrill, R. (1968). Waves of spatial diffusion. *Journal of Regional Science, 8,* 1-18.

Morrill, R. (1970). Shape of diffusion in space and time. *Economic Geography, 46,* 259-268.

Morrill, R. (1974). *Spatial organization of society* (2nd ed.). Belmont: Duxbury.

Morrill, R. (1985). The diffusion of the use of tractors again. *Geographical Analysis, 17,* 88-94.

Morrill, R., & Angulo, J. (1981). Multivariate analysis of the role of school attendance in the introduction of variola minor into the household. *Social Science and Medicine, 15,* 479-487.

Morrill, R., & Manninen, D. (1975). Critical parameters of spatial diffusion processes. *Economic Geography, 51,* 269-277.

Morrill, R., & Pitts, F. (1967). Marriage, migration and the mean information field. *Annals of the Association of American Geographers, 57,* 401-422.

Oakey, R. P. (1984). *High technology small firms: Innovation and regional development in Britain and the United States.* New York: St. Martin's.

Odland, J. (1988). *Spatial autocorrelation.* Newbury Park, CA: Sage.

Pederson, P. (1970). Innovation diffusion within and between national urban systems. *Geographical Analysis, 2,* 203-254.

Pielou, F. (1969). *An introduction to mathematical ecology.* New York: John Wiley.

Pitts, F. (1963). Problems in the computer simulation of diffusion. *Papers, Regional Science Association, 11,* 111-119.

Pred, A. (1971). Large city interdependence and the pre-electronic diffusion of innovations in the U.S. *Geographical Analysis, 3,* 165-181.

Rapaport, A. (1951). Nets and distance bias. *Bulletin of Mathematical Biophysics, 13,* 85-91.

Ricciardi, L. M. (1977). *Diffusion processes and related topics in biology.* New York: Springer.

Rogers, E. M. (1983). *Diffusion of innovations* (3rd ed.). New York: Free Press.

Rogers, E. M., & Havens, E. (1962). Rejoinder to Griliches: Another false dichotomy. *Rural Sociology, 27,* 330-332.

Rogers, E. M., & Shoemaker, F. F. (1971). *Communication of innovations: A cross-cultural approach.* New York: Free Press.

Sahel, D. (1981). *Patterns of technological innovation.* New York: Addison-Wesley.

Sauer, C. O. (1952). *Agricultural origins and dispersals.* New York: American Geographical Society.

Sauer, C. O., & Brand, D. (1930). Pueblo sites in southeast Arizona. *University of California Publications in Geography, 3,* 415-448.

Schumpeter, J. A. (1934). *The theory of economic development.* Cambridge: Harvard University Press.

Shreve, S. R., Lecaczky, J. P., & Gaver, D. P. (1984). Optimal consumption for diffusions with absorbing and reflecting barriers. *SIAM Journal on Control and Optimization, 22,* 55-75.

Soete, L., & Turner, R. (1984). Technological diffusion and the rate of technical change. *Economic Journal, 94,* 612-623.
Soja, E. W. (1968). *The geography of modernization in Kenya: A spatial analysis of social, economic and political change.* Syracuse: Syracuse University Press.
Stroock, D. W., & Varadhan, S.R.S. (1979). *Multidimension diffusion processes.* Berlin: Springer.
Thomas, R. W. (1976). *An introduction to quadrat analysis.* London: Institue of British Geographers, CATMOG No. 12.
Thrall, G. I. (1984). Geoinvestment: The interdependence among space, market size and political turmoil in attracting foreign direct investment. *Conflict Management and Peace Science, 8,* 17-48.
Thrall, G. I. (1987). *Land use and urban form: The consumption theory of land rent.* London: Methuen.
Tinline, R. (1971). Linear operators in diffusion research. In M. Chisholm et al., (Eds.), *Regional forecasting,* (pp. 71-91). London: Butterworth.
Tobler, W. (1965). Computation of the correspondence of geographical patterns. *Papers, Regional Science Association, 15,* 131-139.
Veblen, T. (1899). *The theory of the leisure class.* New York: Macmillan.
Webb, W. P. (1927). *The great plains.* Boston: Ginn.
Webber, M. (1972). *The impact of uncertainty on location.* Cambridge: MIT Press.
Webber, M. (1984). *Industrial location.* Newbury Park, CA: Sage.
Webber, M., & Joseph, A. E. (1978). Spatial diffusion processes 1. *Environment and Planning, A 10,* 651-665.
Webber, M., & Joseph, A. E. (1979). Spatial diffusion processes 2. *Environment and Planning, A 11,* 335-347.
Yapa, L. (1975). Analytic alternatives to the Monte Carlo simulation of spatial diffusion. *Annals of the Association of American Geographers, 65,* 163-176.
Yuill, R. (1964). *A simulation of barrier effects in spatial diffusion.* ONR Spatial Diffusion Study, Northwestern University. Department of Geography.

ABOUT THE AUTHORS

RICHARD MORRILL is a native of Los Angeles. However, apart from serving as Visiting Professor at University of Chicago, the University of Washington is his academic home. He received his Ph.D. there in 1959, remained to become Professor of Geography and Environmental Studies, and to serve as Chairman of its Department of Geography. Torsten Hägerstrand's 1959 lectures at University of Washington were an inspiration for his subsequent research. Morrill's multidimensional works range between being of the highly abstract kind, such as his *Spatial Organization of the Landscapes,* to contributions to public policy analysis, such as his *Political Redistricting and Geographic Theory, The Geography of Poverty,* to being applied, such as the Chicago Regional Hospital Study, which he directed while on leave at University of Chicago. The research theme common throughout these eclectic works is his interest in the foundations of the theory of spatial relationships and behavior of persons in space. He is an active proponent of the science of geography being applied to issues of social equity. He is a 1983 recipient of a Guggenheim Fellowship and is past president of the Association of American Geographers.

GARY L. GAILE is a native of Ohio, but received his university education, culminating with his Ph.D. in 1976, from the University of California at Los Angeles. He was Assistant Professor at Northwestern University prior to his present positions as Associate Professor at University of Colorado, Boulder, and Visiting Scholar with Harvard University and its African Development Program in Kenya. His research interests include spatial mathematical and computational methods, basic research on diffusion theory, and applications of diffusion methods to Third World development. He is coeditor of the book *Spatial Statistics and Models*, and coauthor of the book *Directional Statistics.*

GRANT IAN THRALL too is a Californian, but left his native San Gabriel for Ohio State University, where he completed an M.A. in Economics and in 1975 a Ph.D. in Geography. He has held faculty appointments at McMaster University in Canada, and at SUNY/Buffalo, where he taught both in the Department of Economics and Department of Geography. In 1985 he became Professor of Geography at University of Florida. He has published numerous articles, mostly in geography and regional science journals. His research interests have focused upon the spatial form of cities, housing, the public sector, and human welfare. His theory of urban spatial structure is published as *Land Use and Urban Form: The Consumption Theory of Land Rent*. Locally, he is Vice-Chairman of the Gainesville and Alachua County Greenbelt Advisory Task Force. He has served as Chairman of the Association of American Geographers' Mathematical Models and Quantitative Methods Specialty Group. In 1982 he proposed to Sage Publications that they publish a collection of small books that became this *Scientific Geography Series*.

NOTES

NOTES